Dessein H.

VERSAILLES
SCIENCE & SPLENDOUR

N°1.

Se vend a Paris chez Basan et Poignant rue et Hotel Serpente.

VERSAILLES

SCIENCE & SPLENDOUR

EDITED BY ANNA FERRARI

SCALA

Contents

Directors' Forewords

SIR IAN BLATCHFORD

Versailles: Science and Splendour opens at the Science Museum after many years in the planning. Ever since *Sciences et curiosités à la cour de Versailles* took place at the Palace of Versailles in 2010–11, I have been convinced that it would delight visitors to the Science Museum in London.

Our exhibition breaks new ground and presents the court of Versailles – the seat of the French absolute monarchy – in a new light. It reveals the royal palace, commonly associated with its architecture, gardens and splendour, as a place of serious scientific enquiry and expands our understanding of Versailles and its mystique. Highlighting the relationship between science and power from the late 1600s to the French Revolution, we explore the ways in which the monarchy and the court harnessed science to enhance their prestige and extend France's global influence. From institutions such as the Royal Academy of Sciences to the gardens' engineering, and from the menagerie of rare animals to the king's scientific cabinets, we have brought together extraordinary objects as diverse as an obstetrics teaching model, a stuffed rhinoceros and an exquisite astronomical clock. I hope *Versailles* will inspire visitors to the Science Museum and deepen their knowledge and appreciation of a remarkable palace and of the history of French science.

The Science Museum is the ideal place to present this exhibition in the United Kingdom. It is home to the national collections relating to science and, since its origins in the nineteenth century, it has collected art as integral to an understanding of the history of science. Our galleries and exhibitions display art and science together to demonstrate that science is a cultural practice. *Science City 1550–1800: The Linbury Gallery* examines the birth of scientific culture in London and displays objects from King George III's collection on loan from King's College London. *Versailles: Science and Splendour* focuses on similar themes. It follows a suite of exhibitions exploring the interplay between visual and scientific culture, including *Art of Innovation* (2019–20), *Ancient Greeks: Science & Wisdom* (2021–22) and *Zimingzhong 凝时聚珍: Clockwork Treasures from China's Forbidden City* (2024).

This exhibition would never have happened without the support of Béatrix Saule, who conceived and curated the original exhibition as the Director and General Curator of the Musée national des châteaux de Versailles et de Trianon, and of many staff at the Palace of Versailles itself. I am deeply grateful to Béatrix Saule and to colleagues at the Palace for embracing the idea of bringing the show to London and for sharing their expertise so generously. At the Science Museum, I would like to thank the team who worked tirelessly to organise this exhibition and to edit this book. I extend my warmest thanks to the many lenders, particularly in France, who have allowed precious works from their collections to travel and thereby make our exhibition possible. At the present moment, it is more vital than ever to build bridges with our European neighbours and partners.

Sir Ian Blatchford
Director and Chief Executive
Science Museum Group

CHRISTOPHE LERIBAULT

Versailles is a palace of power. And this power – this is well known – is embodied, even embedded, in the scale of the château, the seemingly unlimited perspectives of the gardens, the luxurious decorations, the refined furnishings, the abundant fountains, the profusion of artworks, but also – and this is less known – in the flourishing of science.

This aspect is less obvious to visitors today, largely because many scientific objects housed at Versailles were scattered during and after the Revolution, stripping the Palace of this important part of its past. The ongoing refurnishing of the Palace cannot bring back the full wealth of scientific objects, instruments and curiosities that it once held.

Nonetheless, in reimagining the 2010–11 exhibition *Sciences et curiosités à la cour de Versailles* at the Palace, the Science Museum has brought together as many artefacts as possible, from Versailles but also from Paris, London, Rouen and Reims. Scattered, they could seem of little importance, the manifestation of a diffuse, haphazard interest in science. Gathered together, they are evidence of the state's consistent attention to scientific topics during the Enlightenment.

Science supported the national ambitions of the kings. Louis XIV, Louis XV and Louis XVI were interested in technologies that benefited the health, economy and administration of the realm, such as medicine, geography and agriculture. Scientific innovations were also key to building the wonders of Versailles, thereby allowing the King of France to dazzle his subjects and assert his sacred authority. Being at the forefront of scientific progress was equally decisive in ensuring the power and influence of the French kingdom in Europe, and all around a world that was better known, but also more colonised, after each expedition. The preparatory map of La Pérouse's voyage is one of the best examples of this blend of scientific, imperialist and colonial interests (Ill. 68).

Louis XV's and Louis XVI's personal passions for science fuelled a general interest in scientific pursuits. Royal patronage was echoed in society by different figures. The portrait of Emilie du Châtelet (Ill. 44) sheds light on one of these prominent figures, and reveals the important role played by women in the scientific world, in spite of their severely restricted opportunities. Madame du Coudray's obstetric teaching model showcases women's innovations in science and education, for the broader good (Ill. 52).

I extend my warmest thanks to Béatrix Saule, former Director and General Curator of the Musée national des châteaux de Versailles et de Trianon, who conceived and curated the original exhibition, and Anna Ferrari who remodelled it for this new presentation. Many thanks, also, to my colleagues who shared their expertise across the Channel, and to all the lenders who allowed this exhibition to be so rich. I commend the Science Museum and its Director and Chief Executive, Sir Ian Blatchford, for his enthusiasm and determination to carry this project forward since 2010. We are delighted by this exchange – a scientific exchange, too – and the new bridge it builds between our institutions and our countries.

Christophe Leribault
President of the Public Establishment of the Palace,
the Museum and the National Estate of Versailles

Acknowledgements

ANNA FERRARI

Versailles: Science and Splendour and its associated publication would not have been possible without the generosity of a great many individuals and institutions in both France and the United Kingdom. I would especially like to thank Béatrix Saule, who originally conceived this exhibition as Director and General Curator of the Musée national des châteaux de Versailles et de Trianon and who shared her expertise with the Science Museum team. At the museum, I am grateful to: Ian Blatchford for his vision and commitment to the project; Jane Desborough, Matthew Howles, Glyn Morgan and Sarah Wood, as well as Surya Bowyer, Rebekah Chitson, Helen Hughes, Rebecca Mellor and Lucy Twisleton in the Curatorial and Interpretation teams; and Gemma Allen, Hannah Evans and Emma Hedderwick in the Exhibitions team. I am also deeply indebted to Ludmilla Jordanova, former Trustee of the Science Museum Group, and Natasha McEnroe, former Keeper of Medicine at the museum, for their invaluable advice and for reading an early draft of this book.

I am extremely grateful to all the institutional and private lenders to the exhibition for their support and generosity. In particular, I would like to thank the team at the Palace of Versailles for their enthusiasm in bringing this exhibition to London: Hélène Delalex, Christophe Leribault, Elsa Mifsud, Catherine Pégard, Silvia Roman Buonsanti and Laurent Salomé. My thanks to: Karine Alexandrian, Catherine Arnold, Andréa Barbe-Hulmann, Charlotte Bellando, Pierre-Emmanuel Biot, Charlotte Brooks, Rémi Cariel, Anne-Laure Carré, Fabienne Casoli, Elisabeth Caude, the Château de Breteuil, Louise Contant, Simon Dade, Frédéric Dassas, Bruno David, Fiona Davison, Christophe Degueurce, Catherine Delot, Sophie Demoy-Derotte, Laurence Des Cars, Louise Devoy, Mélanie Drappier, Laurence Engel, Marie Laure Estignard, Claire Fantino, Guillaume Faroult, Cyrille Foasso, Olivier Gabet, Alice Gamblin, Julie Garel-Grislin, Sophie Guérinot, Douglas Gurr, Lieutenant-colonel Philippe Guyot, Vincent Haegele, Andrea Hart, Maxence Hermant, Jessica Hudson, Olivier Labat, Hélène Lachaier, Brigitte Laude, Olivier Laville, Claire Le Borgne, Christine Lefèvre, Alice Lemaire, Hortense Longequeue, Ophélie Louis, Emilie Maisonneuve, Sarah McDonald, Général Henry de Medlege, Eve Netchine, Anna Nouet, Clément Oury, Agnès Parent, Christophe Pérales, Amandine Postec, Emilie Prud'hom, Paddy Rodgers, Simon Stephens, Véronique Stoll, Molly Tillett, Aleth Tisseau des Escotais and Marianne Tricoire.

Many scholars and curators have contributed to this project and I am immensely grateful for their vital help. I would like to thank the authors of this book who are listed on pp. 124–25, many of whom have also shared their knowledge more widely, as well as the advisory panel to this exhibition, Patricia Fara, Jean-François Gauvin, Ludmilla Jordanova, Sir Paul Nurse and Emma Spary. I would also like to thank the scholars who have offered their advice over the years and have enriched the project: Michael Bycroft, Sarah Easterby-Smith, Louisiane Ferlier, Robert Fox, Helen Jacobsen, Colin Jones, Bruno Maquart, Alice Minter, Keith Moore, Gregory Radick, Catriona Seth, Thomas Stammers, Anthony Turner, Stéphane Van Damme, Simon Werrett and Gabriel Wick.

For the book, I would like to thank: Wendy Burford in the SMG Publications team; Andrew Tunnard for coordinating the photography of Science Museum objects; Lily Hayward and Holly Palmer for assembling the images; Oliver Craske at Scala and our editor Linda Schofield; Peter Dawson at Grade Design for the elegant design of this book; and Rae Walter and Robert Anderson in association with First Edition Translations Ltd for translations from the French and for editing, respectively.

Many colleagues from across SMG have been instrumental in bringing this exhibition together. I would like to thank: Jessica Crann, Sophie Croft, Richard Horton and Emily Yates for conserving objects for display; Laurence Deane, Marianne Dear, Sonia D'Orsi, Kenny Foot and Becky Jarvis-Stiggants in the Design team for their beautiful exhibition designs and graphics; David Dewhurst and Irina Tsokova in the New Media team for their support; Ruth Clapham, Lucy Findley, Victoria Sall and Nicole Simoes da Silva in Registration for coordinating the loans; Elizabeth Campbell, Carole Pevny and Kit Webb for delivering the exhibition; and Chloë Abley, William Dave, Laura Nebout and Jack Stanley for promoting it.

Such a project would not have been possible without the support of colleagues from the Collections team: Sarah Bond, Alison Boyle, Jessica Bradford, Gabrielle Bryan-Quamina, Rupert Cole, Katie Dabin, Richard Dunn, Stewart Emmens, Selina Hurley, Harriet Jackson, Helen Rayner, Ben Russell, Prabha Shah, Kathleen Walker-Meikle and Nick Wyatt.

LENDERS

Bibliothèque de l'Observatoire de Paris
Bibliothèque municipale de Versailles
Bibliothèque nationale de France, Paris
Ecole nationale vétérinaire d'Alfort – Musée Fragonard, Maisons-Alfort
Musée de l'Armée, Paris
Musée des Arts et Métiers – CNAM, Paris
Musée des Beaux-Arts, Reims
Musée d'Histoire de la Médecine | Université Paris Cité
Musée du Louvre, Paris
Musée Flaubert et d'histoire de la médecine, Réunion des musées, Métropole Rouen Normandie
Musée Lambinet, ville de Versailles
Musée national de la Marine, Paris
Musée national des châteaux de Malmaison et Bois-Préau, Rueil-Malmaison
Musée national des châteaux de Versailles et de Trianon
Muséum national d'Histoire naturelle, Paris
National Maritime Museum, Greenwich, London
Private collection Château de Breteuil – France
RHS Lindley Collections, London
The Trustees of the Natural History Museum, London

Curator's Note

ANNA FERRARI

Versailles: Science and Splendour explores the relation between science and power at the court of Versailles between the 1660s and the beginning of the French Revolution in 1789. The genesis of the show at the Science Museum was the exhibition *Sciences et curiosités à la cour de Versailles* (Sciences and Curiosities at the Court of Versailles) that opened at Versailles in 2010. Curated by Béatrix Saule, who was then Director and General Curator of the Musée national des châteaux de Versailles et de Trianon, it broke new ground and presented the palace as a site of science. In bringing this exhibition to the Science Museum 14 years later, we have sought to build on the success of this show and adapt it to our museum dedicated to science, technology, engineering, mathematics and medicine – a very different context to the Baroque grandeur of Versailles. In doing so, we have had three considerations in mind.

First, the curatorial team wished to explore the relation between science and power at the French court of Versailles, widening the focus from the three kings, Louis XIV, Louis XV and Louis XVI, while also bringing the history of science to the fore in a period when the practice of what we now call 'science' changed significantly, with direct observation and experimentation gaining new currency.[1] In the mid-seventeenth century, the foundation of scientific institutions across Europe, such as the Académie royale des sciences (Royal Academy of Sciences) in Paris in 1666, created a framework linking scientific research with political power and state ambitions. Jean-Baptiste Colbert, Comptroller-General of Finances under Louis XIV, established a network of institutions, which included the Royal Academy of Sciences and the Observatoire (Paris Observatory), to organise and centralise knowledge, and also serve the monarchy (Ill. 1). These bodies were part of a wider scientific infrastructure, which continued to develop in the eighteenth century and became strategically important for ruling and administering the French kingdom. They were also central to imperial expansion, becoming part of what has been called the 'scientifico-colonial machine'.[2]

Second, we aimed to highlight how science and empire were connected at Versailles. France's imperial reach enabled the court to become a centre for the serious study of plants and animals from around the world, while science and technology were used strategically by all three kings for defence and to develop imperial power. At the same time as Colbert founded scientific institutions, he also established the Compagnie des Indes occidentales (French West Indies Company) to manage French territories in the Americas, and the Compagnie

ILL. 1
Le Cabinet de M. Le Clerc (The Cabinet of Mr Le Clerc), possibly showing the Academy's collection of instruments or an idealised scientific cabinet, by Sébastien Le Clerc the Elder, *c.*1711
Science Museum Group, London. Object no. 1948-336

des Indes orientales (French East India Company) to colonise Madagascar, as well as the Ile Bourbon (Réunion), and to set up trading posts in India, challenging Portuguese, English and Dutch supremacy in the Indian Ocean. Both companies were founded in 1664 and demonstrated French ambition to build an empire. By the end of Louis XIV's reign, France had a presence in North America (in Canada and Louisiana – named in honour of the king), the West Indies, in the Indian Ocean with the Ile Bourbon, and was developing trading posts in the Levant, India and West Africa. These territories were contested but nonetheless extended French reach. Versailles became a hub, receiving animals and plants from the four corners of the world that were studied by French naturalists.

Finally, as we considered the connections between science and power and the importance of institutions such as the Royal Academy of Sciences, we wanted to shed light on the roles women played, even though academies typically excluded women.[3] We seek to explain why women are less present than men in this exhibition: there were limited opportunities for intellectual women in France in the seventeenth and eighteenth centuries, with fewer women than men learning to read. Under Louis XIV only 14 per cent of French women could read compared to 29 per cent of men, and by the end of Louis XVI's reign still only 27 per cent of women could read compared to 47 per cent of men.[4] Yet some women managed to play significant roles and shaped natural philosophy during this period. The aristocratic Emilie du Châtelet is one of the often-cited exceptions who succeeded as an influential philosopher, mathematician and physicist in the 1730s and 1740s. Another notable woman is Angélique Marguerite Le Boursier du Coudray, who

ILL. 2
Portrait of Sophie-Philippine-Elisabeth-Justine de France, known as Madame Sophie, by Lié Louis Périn-Salbreux, 1776
Musée des Beaux-Arts, Reims. Object no. 923.3.1

sought to improve both the education of midwives and birth outcomes across France in the second half of the eighteenth century. Identifying objects that would enable us to evoke their work sometimes proved challenging. Madeleine Basseporte was an important figure for us to represent, if only with one watercolour of a salvia plant testifying to her study and scientific illustration of botanical specimens. Born in Paris in 1701, she took lessons from Claude Aubriet, the official draughtsman at the Jardin du Roi (King's Garden), before eventually succeeding him to this prestigious post in 1741. Despite her long career, there are relatively few identified drawings, perhaps in part because Aubriet signed many of her works and sold them as his during his lifetime.[5]

Documentation concerning women is uneven and relatively little is known about their interests, even in the case of royal figures, such as the daughters of Louis XV, Madame Adélaïde and Madame Sophie. One historical source published in the early twentieth century dismissively describes Madame Adélaïde's broad interests as superficial: 'She wished to play all musical instruments [...]; she learnt English, Italian, high mathematics, clockmaking, without deepening her knowledge.'[6] However, in 1776 the painter Lié Louis Périn-Salbreux presented them in a different light, sitting at their desks engaged in intellectual activities, Sophie reading and Adélaïde writing (Ill. 2, 3). The walls are lined with bound volumes and, while libraries were typical features of royal apartments, we know from Madame Adélaïde's library catalogue that she owned an impressive 11,000 books. The volumes in her collection spanned a broad range of subjects from prayer books to philosophy and law, history and geography, poetry and literature,

ILL. 3
Portrait of Marie-Adélaïde de France, known as Madame Adélaïde, by Lié Louis Périn-Salbreux, 1776
Musée national des châteaux de Versailles et de Trianon. Object no. MV 9085

as well as sciences and arts, revealing her varied interests. While even less seems to be known about Madame Sophie, a telescope in the collection of the Musée de la Marine in Paris, bearing the inscription 'Made by Madame Sophie de France', also hints at more than a passing interest in scientific subjects.[7]

The resulting exhibition and book weave together these stories to show the French monarchy harnessing science to heighten its prestige and extend France's global influence. Both touch on many facets of seventeenth- and eighteenth-century life, as is evident in our multidisciplinary approach to this rich moment in the history of science.

NOTES

1 It was then more often referred to as 'natural philosophy', a term that had been used since antiquity to refer to a broad field of enquiry aiming to explain and investigate natural phenomena, and which was associated with philosophy and religion. The English term 'scientist' is of nineteenth-century origin and is sometimes used for convenience in this book. The more common use of the French word 'savant' draws attention to the importance of knowledge and avoids distinguishing between disciplines that were only defined in the nineteenth century.
2 McClellan and Regourd 2000.
3 The notable exception was the Accademia delle Scienze dell'Istituto di Bologna (Academy of Sciences of the Institute of Bologna), which accepted women.
4 Jones 2003, p. 185.
5 Gelbart 2021, p. 117.
6 Stryienski 1911, p. 61.
7 Musée de la Marine, object no. 15 NA 17 D.

Introduction

BÉATRIX SAULE

Once a country residence devoted to hunting and festivities, in 1682 the Palace of Versailles became the seat of power and the official residence of the French court (Ill. 4). Three kings ruled from Versailles: Louis XIV until his death in 1715, then Louis XV from his coming of age in 1722 until 1774, and finally Louis XVI until the Revolution of 1789. It was the end of the classical period: the Age of Enlightenment, the century of triumphant absolutism and of the *philosophes*, the time of the Royal Academy of Sciences and the *Encyclopédie*.[1]

Court society at Versailles was then a whole world in itself. All ranks and occupations were represented, from members of the royal family and ministers of state to grooms and kitchen boys. A census carried out at the beginning of the eighteenth century counted some three thousand permanent residents of the palace and its related buildings. An important hub where many came to meet or conduct business, it is little wonder that Versailles attracted prominent 'savants' (scientists), including natural philosophers.

At that time the estate was of a considerable size, 8,000 hectares compared with 800 today. In addition to the palace itself, the walls surrounded the Trianon estate with its two residences, the gardens and parks, and numerous buildings, some of which, such as the two stable blocks, were located in the town (Ill. 5). In addition, the estate of Versailles included more private dwellings, such as Marly, the favourite residence of Louis XIV, and the so-called *petits châteaux* of Choisy and La Muette, the preferred retreats of Louis XV.

THE SCIENCES: A STATE AFFAIR

Influenced by Jean-Baptiste Colbert, one of his ministers from 1661 onwards, Louis XIV realised the crucial importance of scientific research. The founding of the Royal Academy of Sciences in 1666 established a new contract between the monarchy and the savants whose work was to be of benefit to the realm. The following year the establishment of the Paris Observatory, financed by the royal treasury, gave new momentum to the search for a solution to the problem of longitude, which persisted throughout the eighteenth century and whose strategic significance can be compared to that of atomic research in more modern times. This is one of many examples of the growth of a scientific policy.

At Versailles, the supervision of the Academy and consequently of official scientific activities was overseen by a minister of the crown. This person played a part in the nomination of academicians, was involved in the establishment of many schools and academies of science, instigated or supported *enquêtes* (enquiries) throughout France and expeditions to distant lands, and granted pensions or awards to savants or recommended them for particular positions.

The main areas of royal involvement in the promotion of scientific advances were those that required substantial funding and would also achieve the monarch's objectives. Under the influence of the *Encyclopédistes* these objectives gradually shifted away from the enhancement of the king's prestige and the practical research of Colbert's time towards the morally useful and beneficial to the public good. Hitherto, the applied

ILL. 4
A Stag Hunt at Versailles, attributed to Jean-Baptiste Martin, *c.*1700
Royal Collection Trust, London. Object no. RCIN 406958

sciences had been favoured over fundamental research and, just as in the present day, the disciplines preferred by the monarchy were those offering military, economic and industrial advantages, including astronomy for navigation, geometry and chemistry for the artillery, geodesy and cartography for mapping the French kingdom for cadastral purposes, medicine and pharmacy for public health, and agronomy to combat famine.

SAVANTS AT COURT

At the pinnacle of this court society the king was involved in the field of science as both sovereign and private individual. Louis XIV saw himself as the protector of the sciences – as well as of the arts – but without practising them. By contrast, Louis XV and Louis XVI were true connoisseurs who carried out observations and experiments. As for the princes, only the Enfants de France (the sons and grandsons of the sovereign) were brought up at court and had access to a 'house' with their own personal scientists. For the 'health officers' – the doctors, surgeons and apothecaries – the position represented the highest honour of their career. Reputation also influenced the selection of the masters recruited by the *gouverneurs* and tutors of the princes: geometers, geographers, astronomers and military engineers were appointed for the period of tuition; the post of assistant tutor for the sciences was not created until the mid-eighteenth century. Other scientists came to court when specialist knowledge was required, for instance the chemist Antoine Lavoisier, who was summoned twice. Courtiers were not overlooked: the duc de Chaulnes,

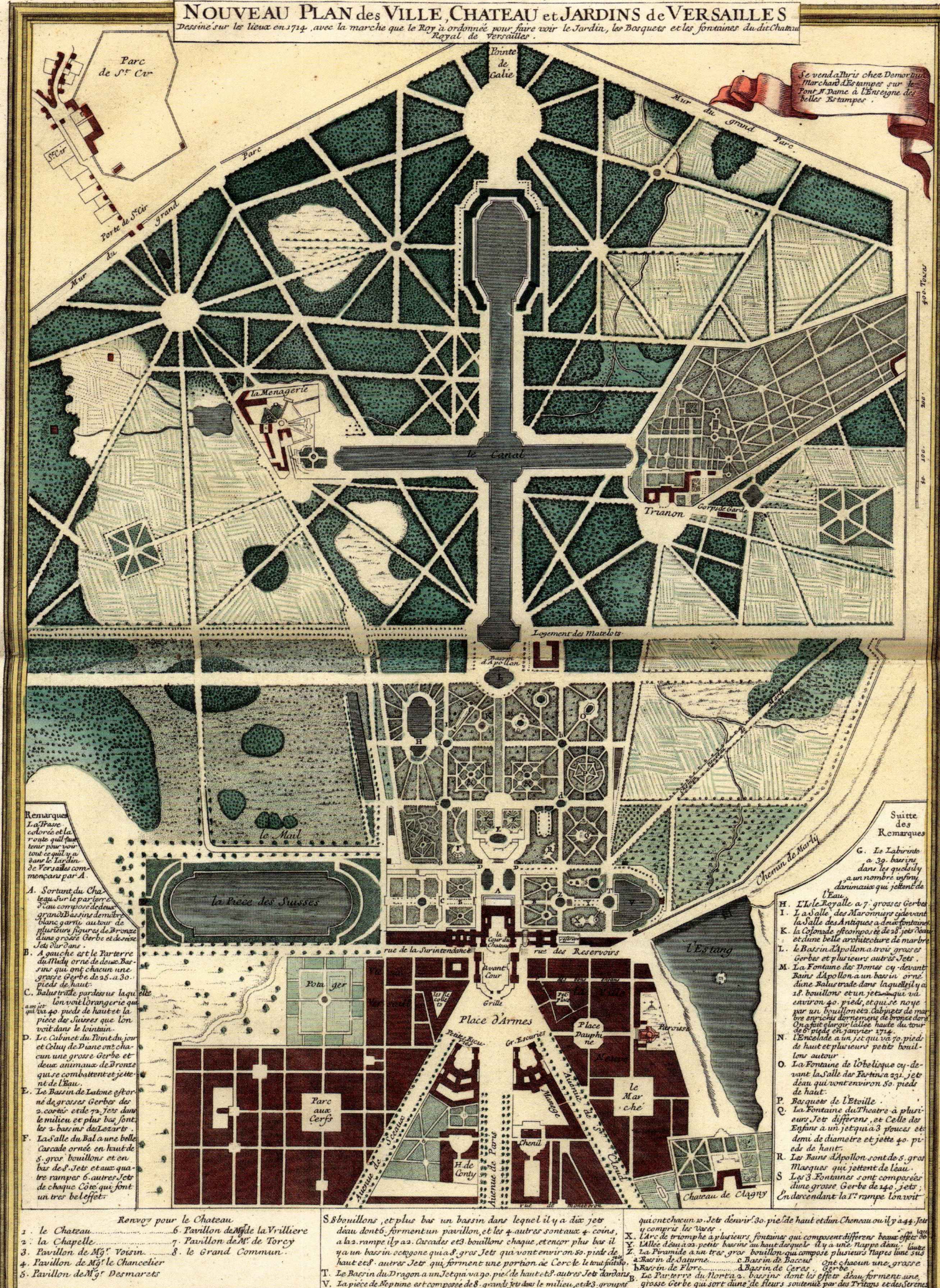

NOUVEAU PLAN des VILLE, CHATEAU et JARDINS de VERSAILLES
Dessiné sur les lieux en 1714, avec la marche que le Roy à ordonnée pour faire voir le Jardin, les Bosquets et les fontaines du dit Chateau Royal de Versailles.
Se vend a Paris chez Demortain Marchand d'Estampes sur le Pont N. Dame à l'Enseigne des belles Estampes.
Parc de St Cyr
Porte de St Cir
Mur du grand Parc
Mur du grand Parc.
Pointe de Galie
la Menagerie
le Canal
Trianon
Corps de Garde
Logement des Matelots
le Mail
la Piece des Suisses
l'Estang
Chemin de Marly
Potager
Place d'Armes
Place Dauphine
le Marché
Parc aux Cerfs
H. de Conty
Chenil
Chateau de Clagny
Avenue de Paris
Avenue de St Cloud
rue de la Surintendance
rue des Reservoirs
Remarques
La Trasse colorée et la route qu'il faut tenir pour voir tout ce qu'il y a dans le Jardin de Versailles commençans par A.
A. Sortant du Chateau Sur le parterre d'eau composé de deux grands Bassins de marbre blanc garni autour de plusieurs figures de Bronze d'une grosse Gerbe et de seize Jets d'ardans.
B. A gauche est le Parterre du Midy orné de deux Bassins qui ont chacun une grosse Gerbe de 25. a 30. pieds de haut.
C. Balustrade par dessus laquelle l'on voit l'orangerie qui a un jet qui va 40. pieds de haut et la piece des Suisses que l'on voit dans le lointain.
D. Le Cabinet du Point du jour et Celuy de Diane ont chacun une grosse Gerbe et deux animaux de Bronze qui se combattent et jettent de l'Eau.
E. Le Bassin de Latone est orné de grosses Gerbes des 2. costés et de 72. Jets dans le milieu et plus bas sont les 2. bassins des Lezarts.
F. La Salle du Bal a une belle Cascade ornée en haut de 5. gros bouillons et en bas de 8. Jets et aux quatre rampes 6. autres Jets de chaque Côté qui font un tres bel effet.
Suitte des Remarques
G. Le Labirinte a 39. bassins dans les quels il y a un nombre infiny d'animaux qui jettent de l'Eau.
H. L'Isle Royalle a 7. grosses Gerbes.
I. La Salle des Maronniers cy devant la Salle des Antiques a deux fontaines.
K. la Colonade est composée de 28. jets d'eau et d'une belle architecture de marbre.
L. le Bassin d'Apollon a trois grosses Gerbes et plusieurs autres Jets.
M. La Fontaine des Domes cy-devant Bains d'Apollon a un bassin orné d'une Balustrade dans laquelle il y a 18. bouillons et un jet qui va environ 40. pieds, et qui se noye par un bouillon et 2. Cabinets de marbre enrichis d'ornemens de bronze doré. On a fait elargir l'allée haute du tour de 6. pieds en janvier 1714.
N. l'Encelade a un jet qui va 70. pieds de haut et plusieurs petits bouillons autour.
O. La Fontaine de l'obelisque cy-devant la Salle des Festins a 231. jets d'eau qui vont environ 50. pieds de haut.
P. Bosquets de l'Etoille.
Q. La Fontaine du Theatre à plusieurs Jets differens, et Celle des Enfans à un jet qui a 3 pouces et demi de diametre et jette 40. pieds de haut.
R. Les Bains d'Apollon sont de 5. gros Masques qui jettent de l'eau.
S. Les 3. Fontaines sont composées d'une grosse Gerbe de 140. jets; En descendant la 1re rampe l'on voit 8 bouillons, et plus bas un bassin dans lequel il y a dix jets d'eau dont 6. forment un pavillon, et les 4. autres sont aux 4. coins, a la 2. rampe il y a 2. Cascades et 3. bouillons chaque, et encor plus bas il y a un bassin octogone qui a 8. gros Jets qui vont environ 50. pieds de haut et 8. autres Jets qui forment une portion de Cercle le tout fait 80. jets
T. Le Bassin du Dragon a un Jet qui va 90. pies de haut et 8. autres Jets dardans
V. La piece de Neptune est composée de 8. grands jets dans le milieu, et de 3. groupes qui ont chacun 10. Jets d'envir. 30. pies de haut et d'un Cheneau ou il y a 44. Jets y compris les Vases.
X. l'Arc de triomphe à plusieurs fontaines qui composent differens beaux effets de l'eau
Y. l'Allée d'eau a 22. petits bassins au haut desquels il y a une Nappe d'eau.
Z. La Piramide a un tres gros bouillon qui compose plusieurs Napes l'une sus l'autre
a. Bassin de Saturne
b. Bassin de Flore
c. Bassin de Baccus
d. Bassin de Ceres.
Ont chacun une grosse Gerbe
B. Le Parterre du Norta 2. bassins dont les effets d'eau forment une grosse Gerbe qui sort d'une Couronne de fleurs soutenue par des Tritons et des Serenes
Renvoy pour le Chateau
1. le Chateau
2. la Chapelle
3. Pavillon de Mgr. Voisin
4. Pavillon de Mgr le Chancelier
5. Pavillon de Mgr Desmarets
6. Pavillon de Mgr de la Vrilliere
7. Pavillon de Mr. de Torcy
8. le Grand Commun.

ILL. 5
Plan of the city, palace and gardens of Versailles as they appeared in 1714, a plate from *Les Plans, profils et élévations des ville et château de Versailles* ... (Plans, Profiles and Elevations of the City and Palace of Versailles ...), published by Gilles Demortain (Paris, 1715)
Bibliothèque nationale de France, département Arsenal, Paris. Object no. EST-39

the most scientifically minded of the courtiers, rubbed shoulders with the naturalist Georges-Louis Leclerc, comte de Buffon ('the most courtly of the scientists'), and Emilie du Châtelet, the eminent translator of Newton. Great enlightened aristocrats were chosen to occupy the ten statutory posts of honorary members of the Academy of Sciences.

The presence of these savants at court attracted others, notably Benjamin Franklin, who went to Versailles on a diplomatic mission and discussed his theories on the nature of electricity and the lightning conductor with the abbé Nollet, the physics tutor of the Enfants de France; or François Quesnay, doctor to Louis XV, who welcomed *Encyclopédistes*, including Denis Diderot and Jean-Baptiste le Rond d'Alembert, in his rooms situated at Madame de Pompadour's apartments. So, even at the time of the Enlightenment, the barrier separating academic and courtly circles appears to have been far from watertight. They met at the court of Versailles and in the salons of Paris (Ill. 6).

VERSAILLES, A PLACE WHERE SCIENCE AND TECHNIQUES WERE APPLIED

The unprecedented scale of the construction at Versailles represented a real challenge. In addition to the traditional skills, the works demanded new scientific and technical knowledge. Even more so than for the buildings, new solutions were required for the gardens and parks. This was the greatest challenge of the first and most glorious part of Louis XIV's reign, from 1661 to 1690. During this period the landscape had to be thoroughly remodelled in order to create the gardens and parks and to supply the water for the fountains. Before the work was carried out, the land had to be studied through surveying and levelling. New instruments and calculations taking account of the curvature of the earth were developed by the astronomer Jean Picard. His telescopic sights made it possible to measure angles and levels. At the same time, the landscape architect André Le Nôtre designed the expansive

ILL. 6
Lecture de la Tragédie de L'Orphelin de la Chine *de Voltaire dans le Salon de Madame Geoffrin* (The Reading of Voltaire's Tragedy *The Orphan of China* in Madame Geoffrin's Salon), by Anicet Charles Gabriel Lemonnier, 1812
Musée national des châteaux de Malmaison et Bois-Préau, Rueil-Malmaison. Object no. M.M.59.3.1

perspective views across the estate, achieving his visual effects by applying the principles of geometric optics in an original way.

To provide Louis XIV with the pleasure of a fountain display, about 9,500 cubic metres of water were needed for a spectacle lasting two and a half hours – savants and engineers undertook the first calculations of outflow and studies of the friction and resistance of the materials for the piping. The resulting technological advance was that earthenware, wood or lead pipes were replaced by cast-iron pipework consisting of standardised elements that fitted together without the need for welding. The nozzles that shaped the jets of water were perfected and, following the advice of the Dutch academician Christiaan Huygens, the water consumption was reduced. As the system functioned by gravity, it required not only a large quantity of water but also a raised water source. A network of pumps, aqueducts and reservoirs was created, including the famous Marly Machine, which although gigantic ultimately yielded a disappointing supply of water.

VERSAILLES, A PLACE OF SCIENTIFIC RESEARCH

The pleasure grounds, including areas such as the gardens and the menagerie, were also places where scientific research expanded in the fields of zoology, botany and agronomy.

Throughout Louis XIV's reign the Versailles menagerie was filled with birds, exotic cats and

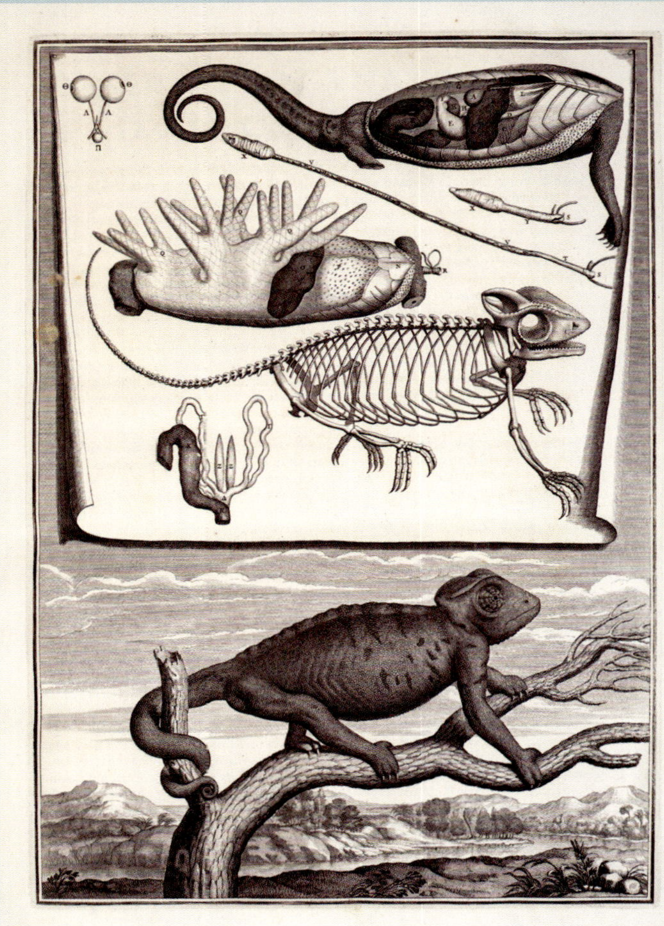

ILL. 7
A chameleon from the Versailles menagerie, a plate from *Mémoires pour servir à l'histoire naturelle des animaux* (Memoirs for a Natural History of Animals), by Claude Perrault (Paris, 1671)
The Trustees of the Natural History Museum, London. Object no. 000398135

other mammals originating from Canada, South America, India, Africa and the Near East, ordered by the governors of the French colonies and transported by the ships of the French East India Company and the French navy. These animals provided the savants with subjects for study. The doctor Claude Perrault carried out dissections that are precisely depicted in the plates of his *Histoire naturelle des animaux* (Natural History of Animals) (Ill. 7). These dissections sometimes took place on-site: in 1681 Louis XIV witnessed the dissection of a Congo elephant and a crocodile. Later, François Gigot de La Peyronie, Louis XV's chief surgeon, benefited from this royal zoo, as did Buffon, the director of the King's Garden in Paris (the future Jardin des Plantes, now both a botanical garden and zoo), who described Louis XV's rhinoceros in detail. Between 1749 and 1751, Louis XV collected a menagerie of domestic animals at Trianon for both his own amusement and for practical use. It brought together farm animals – cows, sheep and hens – of different breeds, in order to cross-breed them and improve the species.

Created to provide food for the royal tables, the 9-hectare Potager du Roi (King's Kitchen Garden), close to the palace, was also the site of experiments under the direction of Jean-Baptiste de La Quintinie. Even more than the wonders he performed in growing melons, peaches, pears and peas, his true innovations concerned the cultivation of asparagus plants and fig trees and the use of many expensive cloches and glass frames. In the eighteenth century the Le Normand family acclimatised coffee plants and pineapples there.

In Louis XV's botanical garden Claude Richard, a horticulturalist trained at the Jacobite court of James II, exiled King of England, at Saint-Germain-en-Laye, together with his son Antoine, established the largest botanical collection in Europe, which contained some four thousand varieties. Over more than 30 years plants arrived from every part of the world, brought back by travelling botanists or exchanged with foreign savants. These new plants were acclimatised in three different parts of the garden – the fruit garden, the flower garden and the botanical garden – which were furnished with heated glasshouses and ponds for the aquatic plants.

With a few rare exceptions, the court was not strictly a place where medical experiments were carried out, but it attracted the best practitioners and those who promoted new techniques and thus contributed to the advancement of their discipline. Moreover, kings and princes gave a considerable boost to science and inspired confidence when they submitted their sacred bodies and royal blood to medical treatment. The most notable medical events were the operation on Louis XIV's fistula in 1686 (for which the king's surgeon, Félix de Tassy, created new surgical instruments), and the inoculation against smallpox of Louis XVI and his brothers in 1774 within weeks of Louis XV's death from the disease.

VERSAILLES, A PLACE OF SCIENCE TEACHING

The education of the royal princes favoured scientific disciplines such as geometry for the design of fortifications and artillery, geography for the analysis of maps and military plans, and astronomy, which was considered worthy of kings. In the mid-eighteenth century the experimental sciences (physics and chemistry) really found their place at Versailles. Zoology and botany were taught during walks at Trianon. Then mathematics broke free from the confines of castrametation (the art of choosing and laying out the site of a camp stronghold), and the navy became the object of a whole separate science. Alongside scientific instruments (such as globes and mathematics sets) and treatises dedicated to the young princes, the first teaching instruments appeared, with those of the abbé Nollet at the forefront.

VERSAILLES, A PLACE OF SCIENTIFIC PRACTICE

Louis XV was among the best-educated princes of his time, and his knowledge was of a predominantly scientific nature. From the age of seven he was passionate about geography and cartography, and at 11 he discovered astronomy. In addition to his core education, he touched upon anatomy and surgery, and from medicine he went on to study botany. He read a great deal, assiduously consulted the maps in his geography gallery, attended dissections, studied plants at Trianon and observed the heavens, following every astronomical event. In the park of his *petit château* of La Muette a cabinet of optics and physics housed the largest telescope in existence at the time (see frontispiece).

Mistrustful of some philosophers, the sovereign sought the company of savants. His collections of art and precious objects no longer consisted of paintings, sculptures and mosaics in semi-precious stone like those of Louis XIV, but of scientific instruments. Conceived by the best minds and built by the greatest craftspeople, these miracles of ingenuity and beauty were a spectacular expression of the level of knowledge achieved at the time.

In addition to his personal interest, Louis XVI's scientific and technical practices reveal his wish to raise the military, economic and industrial power of the realm to the highest rank in Europe. The panelling he ordered for the decoration of one of his most private rooms, which could be reached only via his bedroom, still bears witness to this (Ill. 8). An expert in naval matters, Louis XVI surrounded himself with models and plans for the building of ships and views of the French ports, in particular Cherbourg, whose famous breakwater was one of the great ambitions of his reign. Also an expert in cartography, the king corrected the maps himself. Having followed the Pacific voyages of Captain James Cook and their tragic end, he decided on and participated in the preparation of the scientific expedition of Jean-François de Galaup, comte de La Pérouse. In his private apartments spanning several floors, Louis XVI had at his disposal ten laboratories, workshops and libraries, including a cabinet of chemistry, a gallery of physics in which he carried out experiments with electricity, a cabinet of artillery for studying the latest rifles and cannons, two rooms to house his five lathes, and a forge next to the locksmith's and carpentry workshops.

VERSAILLES, A PLACE FOR DEMONSTRATIONS

In the case of an invention, a discovery or an extraordinary construction such as the Clock of the Creation of the World (see Ill. 40), its presentation to the king or demonstration to the court was the supreme mark of success for its creator, the equivalent of a Nobel Prize. It brought hope of acquisition by the Crown or opportunities for manufacture or offers of capital from both the royal treasury and the private individuals gathered at court. However, as the court was fearful of charlatans, royal permission was not easily obtained. Nevertheless, presentations to the king were very frequent. Although demonstrations before the court were rarer than presentations, they were more akin to a scientific spectacle. They took place in settings that were unusual for this kind of event: at the time of Louis XIV, the Petite Galerie de Mignard was lit entirely by a single candle in front of a burning mirror; under Louis XV, 140 people took part in an electricity experiment in the Hall of Mirrors; and a few years before the Revolution, the forecourt of the palace was the site of the first hot-air balloon flight.

Versailles: Science and Splendour reveals an unexpected side of the palace. What is left on the site today to remind us of how it was? Most of the places still exist, but they have often been transformed and stripped of documents and scientific instruments. These were scattered at the time of the revolutionary sales in 1793–94, when Versailles was emptied of all its furnishings. Fortunately, objects that were considered unique or useful for the education of the people were

ILL. 8
Model of a crane with two jibs and a steelyard, designed by François Jean Bandiery de Laval, 1786–87 (left), with corresponding gilded panelling in Louis XVI's private cabinet at Versailles (right)
Musée des Arts et Métiers – CNAM, Paris (model). Object no. 00076-0000-

removed from the sales and assigned to the collections of successor institutions such as the Bibliothèque nationale, the Bibliothèque municipale de Versailles, the Musée des Arts et Métiers (CNAM) and the Paris Observatory, which have played a large part in the research programme and the creation of the exhibition through their loans. The curatorial department of the Palace of Versailles museum itself has striven to recover the exceptional objects by means of exchanges or permanent loans and is currently endeavouring to identify and acquire engravings, drawings, instruments and other objects from the former royal collections through the art market or from private collectors, in order to restore this part of the collection.

NOTE

1 Commonly referred to as the *Siècle des Lumières*, this was a period that attached great importance to knowledge unfettered by religion, focusing on reason, observation and experimentation.

1. The Royal Academy of Sciences: Painting Science and Power

KATHERINE M. REINHART

A seventeenth-century courtier invited to Versailles would have borne witness to an impressive visual display. From the immense façade to the grandeur of the gardens and the lavish array of artworks, every inch of Versailles was designed by architects and engineers to impress visitors with King Louis XIV's wealth and power. In addition to the gardens, sculptures, tapestries, vases and medals, among the artworks a visitor might have seen the large canvas by Henri Testelin, *Etablissement de l'Académie des sciences et fondation de l'Observatoire, 1666* (The Establishment of the Academy of Sciences and the Foundation of the Observatory, 1666) (Ill. 9). Painted between 1673 and 1681, the image depicts the king meeting the newly formed scientific society of France: the Royal Academy of Sciences.

Founded in 1666, the Academy was part of a wave of institutions created for the pursuit of science (or natural philosophy, as it was more frequently called) in the seventeenth and eighteenth centuries throughout Europe and beyond. Like contemporary institutions such as the Accademia del Cimento in Florence (founded in 1657) and the Royal Society in London (founded in 1660), the French Academy was dedicated to pursuing science collaboratively using tools of the 'new science' including experimentation and observation. However, unlike many of its peer institutions, the Academy, from the beginning, served as an arm of the state. Each Academy member was paid a salary, and all of the group's equipment, materials and lodging were paid for by, and thus in service to, the Crown.

Although, at first glance, Testelin's painting may seem to be a straightforward group portrait, in fact it reveals many of the underlying pursuits, scientific questions and political entanglements of the young institution. The king sits at the centre, denoted by his elaborate clothing and plumed hat, highlighted in vivid red. Next to him, dressed in black, stands his minister Jean-Baptiste Colbert, who established and oversaw the Academy on the king's behalf. Crowded around them are members of the court and founding members of the Academy, such as the Italian astronomer Giovanni Domenico (later Jean-Dominique) Cassini and the Dutch physicist Christiaan Huygens, both prominent philosophers who were recruited by Colbert to France's and the king's service at great expense. Surrounding the men, the room is filled with the objects and instruments of their study.

These objects encompass not only the broad range of the Academy's intellectual pursuits, but also reflect the interests of the king and state. In the left background, animal skeletons evoke the Academy's anatomical investigations, while its explorations in astronomy are referenced by the newly built Paris Observatory

ILL. 9
Etablissement de l'Académie des sciences et fondation de l'Observatoire, 1666 (The Establishment of the Academy of Sciences and the Foundation of the Observatory, 1666), by Henri Testelin after Charles Le Brun, 1673–81
Musée national des châteaux de Versailles et de Trianon. Object no. MV 2074

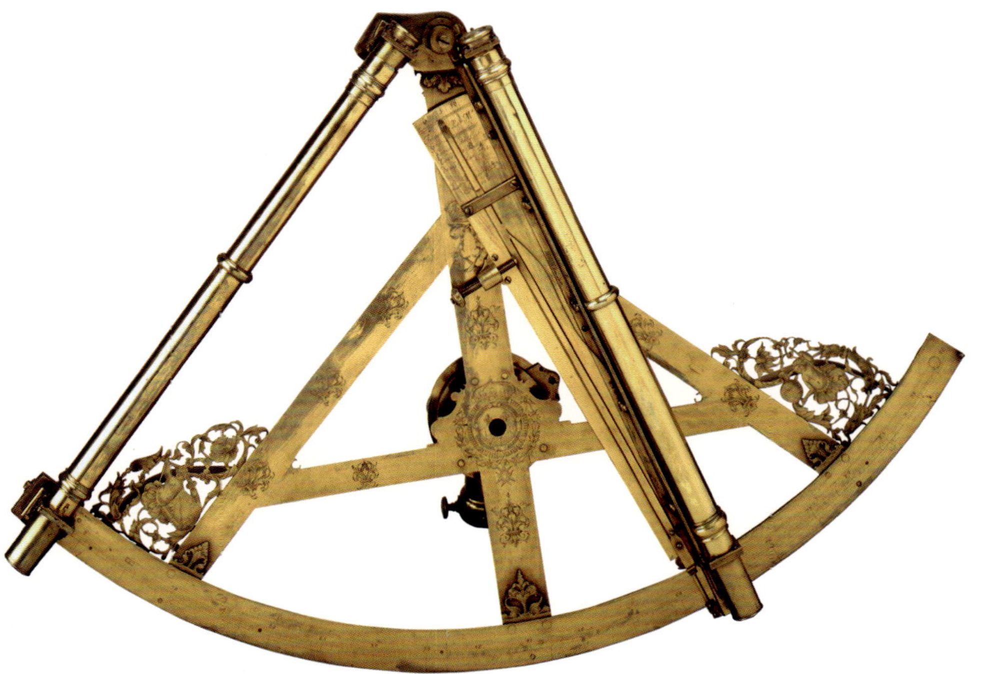

ILL. 10
Astronomical quadrant, divided by Philippe-Claude Le Bas, *c.*1669–76, with a micrometer by Pierre Le Maire II added in *c.*1741
Adler Planetarium, Chicago. Object no. M-186

ILL. 11
Visite de Louis XIV au Jardin du Roi (Louis XIV Visiting the King's Garden), engraved by Sébastien Le Clerc the Elder, *c.*1671
Musée Carnavalet, Histoire de Paris. Object no. G.5220

seen in the centre, as well as instruments such as the armillary sphere in the upper left and the quadrant in the lower right (thought to represent the quadrant that is today in the Adler Planetarium's collection; Ill. 10). The Academy members used these tools in their astronomical enquiries to chart new stars, but also to improve navigation for the king's ships, providing a strategic advantage in war and an economic advantage for trade. Their cartographic work can be seen in the large map on the far right, showing the Canal Royal en Languedoc (later Canal du Midi), a critical seventeenth-century infrastructure achievement connecting the Atlantic to the Mediterranean, and the two globes that frame the scene in the foreground: the terrestrial globe on the left – where the king might survey the geographic reaches of his empire or aspire to new ones – and the celestial globe to the right – another tool to help understand the cosmos and the truth of the heavens. In this way, the Academy's endeavours contributed to the expansion of knowledge, while simultaneously serving the king's interests concerning trade, exploration and military prowess to benefit the French state.

Intriguingly, Testelin's painting combines real objects in an imaginary scene. Many items in the painting, such as Huygens's pendulum clock or Cassini's lunar map, are depictions of specific objects created and used by Academy members (see Ill. 16). Yet, the composition and moment are fictional: the king did not visit the Academy until the 1680s (after the painting was made), nor was the Observatory visible from any of the spaces in which the Academy conducted its work (the King's Library or the Louvre). Despite this, the image of these objects, and their symbolic meanings and direct link to the king, circulated in different visual forms during the Academy's early years. An engraving of an analogous scene

LVDOVICVS · MAGNVS · REX · CHRISTIANISSIMVS

APOLLO PALATINVS
REGIA SCIENT·ACAD·INST·
M·DC·LXVII·
H·ROVSSEL·F·

ILL. 12
Silver medal, by Michel Molart and Jérôme Roussel, 1667
Bibliothèque nationale de France, département des Monnaies, médailles et antiques, Paris. Object no. SR.669

by Sébastien Le Clerc circulated as the frontispiece used in each of the Academy's early books throughout the 1670s (Ill. 11). These symbols of the Academy also circulated in metal, where the Academy's equipment – an animal skeleton, globe, telescope and even Huygens's pendulum clock – appeared on the back of royal medals, often next to allegorical figures, in this case a statuesque Apollo (Ill. 12). The Roman sun god Apollo was associated with Louis XIV, the Sun King, and the figure of the Roman deity, amid the Academy's scientific objects on this silver medal, directly links the two.[1] This relationship is further reinforced by Louis's head in profile on the obverse of the coin. This medal was part of a series created to glorify the king and to impress upon all who viewed it the connection between natural philosophical knowledge and political power – between science and statecraft.

Testelin modelled his canvas on a work by the court painter Charles Le Brun. He designed the original as a tapestry cartoon – a preparatory painting for a tapestry to be included in the *Histoire du roi* cycle.[2] This grand series of large-scale tapestries celebrated the great deeds of Louis XIV's reign. The king had the tapestries displayed throughout Versailles, and they narrated his achievements, from military triumphs to diplomatic visits. For unknown reasons, however, the cartoon of the Academy was never woven into a tapestry. Yet the scene was nonetheless commemorated, only in paint rather than fabric.

If Testelin's magnificent painting did hang at Versailles, as it does today, it would have been seen by members of court and the aristocracy, visiting dignitaries, members of the royal family and, of course, the king himself. Like the other extravagant art objects at Versailles, the image reminded these diverse viewers of the Sun King's monetary and political power, demonstrating that Louis's deep coffers were capable of funding elite savants as well as the expensive equipment and facilities of their enterprise. As a result, the king (and subsequently the French state) was the beneficiary of the knowledge and advancements in industry, agriculture, medicine and warfare generated by this group of natural philosophers.

NOTES

1 Burke 1992, pp. 196–97; Loskoutoff 2016; Jones 1979.
2 Hahn 2010–11, p. 32.

2. Thuret and Clockmaking at the Royal Academy of Sciences

JANE DESBOROUGH

ILL. 13
Spring-driven bracket clock, made by Isaac Thuret, *c*.1665–70
Science Museum Group, London. Object no. 1954-458

Clocks played a vital role in the development of natural philosophy and the story of its formalisation. For this reason, it is not surprising to see a clock featuring symbolically in the Royal Academy of Sciences' representation of itself in Henri Testelin's monumental painting of 1673–81 (see Ill. 9). Similarly, in England the Royal Society's image of itself in Wenceslaus Hollar's frontispiece for Thomas Sprat's *History of the Royal Society* of 1667 also features a clock (Ill. 14).[1]

Within Testelin's painting, located just above the bowing figure of the Academy Secretary, Jean-Baptiste Du Hamel, is a fairly modest-looking clock, certainly by royal standards. The fate and whereabouts of it are unknown, but it bears some resemblance to a clock in the Science Museum's collection (Ill. 13). This clock is signed 'I Thuret Paris'. Its design is in keeping with what is known as *la pendule religieuse*, a specific style of domestic clock made in relatively large numbers by several French makers following the application of the pendulum to clocks in 1657 by the Dutch natural philosopher and inventor, Christiaan Huygens. The pendulum was an effective regulation device for improving the accuracy of clocks, allowing them to keep time to within one minute per day. Before this, clocks could vary as much as half an hour per day. Based on their Dutch predecessors, early *pendules religieuses*, like this one, feature dials consisting of a brass chapter ring bearing Roman numerals on a velvet background with the maker's name on a brass cartouche beneath.[2]

It is significant that this *pendule religieuse* was made by Isaac Thuret. An exceptional craftsman, he was appointed Marchand-Horloger Ordinaire du Roi by 1663, then Horloger Ordinaire du Roi (royal clockmaker) and clockmaker of the Academy and its observatory by 1672, for which he was given a salary of 300 livres per year for maintaining all of their clocks.[3]

Thuret was born around 1630 in Senlis, about 48 kilometres north of Paris. He married Marguerite Helot, whose family were merchants. We assume that he completed the requisite eight-year apprenticeship in clockmaking, given that he was admitted to the Corporation des maîtres horlogers (the Parisian guild of master clockmakers and one of the oldest guilds in Europe) first at Faubourg Saint Germain-des-Prés (then on the fringes of the city) by 1662 and in the city of Paris by 1675. He worked from various addresses in Paris before establishing himself at the prestigious Galeries du Louvre in 1686. He died in the city in 1706, leaving a relatively modest but nevertheless significant estate worth 54,000 livres.[4]

We cannot be certain that the clock in the Testelin painting was made by Thuret, but given his status at court, it can be assumed that he maintained it. It is also

NVLLIVS IN VERBA
CAROLVS
II
OCIETATIS
REGALIS
AVTHOR
&
PATRON
SOCIETATIS PRÆSI
ARTIVM INSTAVRATOR
Evelyn inv D.D.C.
Wencestaus Hollar f. 1667.

ILL. 14
Frontispiece, showing a clock in the left-hand archway, from *The History of the Royal Society of London, for the Improving of Natural Knowledge*, by Thomas Sprat (London, 1667)
Science Museum Group, London. Object no. 1948-398

possible that it was a royal gift from Thuret's most significant patron, Huygens, who can also be seen in the painting standing on the left-hand side below the animal skeletons, with his hand raised, perhaps in a gesture towards the clock.

The date of Thuret and Huygens's first meeting is unknown, but they were certainly working together by 1663–64 and it is through this working relationship and two of Huygens's patents that we can come to know Thuret the inventor-craftsman.[5] The first of Huygens's patents concerns sea clocks, which were designed with the intention of accurately measuring time at sea in the hope of allowing sailors to determine longitude and navigate faster and more safely. A commercially viable marine chronometer was not available until the late eighteenth century, but Huygens was an early experimenter. Better navigation was of vital strategic importance in an age of colonial global expansion and European rivalry for supremacy and control of lucrative trading routes (Ill. 15). Unnecessarily long journeys or damage to ships transporting valuable cargo cost money, so there was a powerful economic and political incentive to solve this problem.[6] From 1663, when Huygens was elected a Fellow of the Royal Society in London, he demonstrated an interest in finding a horological solution to the challenge of determining longitude at sea.[7] In 1665, through Huygens's work on a clock with a remontoire, a device designed to improve accuracy, for which he received a French patent, we see evidence of Thuret's aligned interests.[8] Thuret quickly applied to Jean Chapelain, whom Huygens had appointed to manage the patent in Paris for him, to contribute to the construction of these clocks.[9] This was the cutting edge of horology at the time.

Huygens's work earned him an international reputation and prompted Jean-Baptiste Colbert, Louis XIV's minister, to invite him to Paris. Huygens became a founding member of the Royal Academy of Sciences in 1666, no doubt enticed by the handsome pension and lodgings offered to secure the greatest European scientific minds for France's Academy. Meanwhile, Thuret became the Academy's official clockmaker.

Our second example of patents that brought Thuret and Huygens into contact concerns the latter's spiral balance spring of 1675. Huygens's pursuit of a patent eventually caused a rift between the two men. The spiral balance spring was the equivalent of the pendulum, a regulation device for improving accuracy, but for watches or portable timekeepers. The disagreement began when Huygens commissioned Thuret to make a model from his designs. Thuret later claimed that while doing so he introduced some adaptations and shared these with Huygens. Keen to secure a patent, Huygens showed his model to Colbert in January 1675. Allegedly, when Thuret asked to be named in the patent, Huygens refused, reassuring him instead that he would always be credited verbally and would secure much work from it. Dissatisfied, Thuret approached Colbert himself. In the end, and despite gaining support from elite members of Parisian society, Thuret was forced to rescind his claim. He signed a letter, dated 10 September 1675, in which he pledged his acceptance of Huygens as the exclusive inventor of the spiral balance spring.[10] This must have been a bitter disappointment to such a talented craftsman, but it did not end his horological career. Indeed, he maintained his positions at court and the Academy, continued to make luxury clocks and is remembered as having been the most distinguished Parisian clockmaker of the mid- to late seventeenth century.

Thuret left a significant legacy. His son, Jacques, continued making clocks for wealthy clients, succeeded him as royal clockmaker, and married Louise, daughter of courtier Jean Bérain. It is not known whether Isaac Thuret's daughter, Suzanne, shared the same technical interests as her father and brother. If she did, there would have been few opportunities for her to pursue them. It is known that she married royal tutor Charles-François de Sylvestre, which is indicative of her social status and probably the result of her father's connections.[11] Thuret was an important craftsman for the court and the Academy, a keen inventor and working partner to a natural philosopher. His career exemplifies the range of people that drove early science at the Royal Academy of Sciences, including in areas that benefited the French monarchy at Versailles.

NOTES

1 Hunter 2017.
2 Plomp 2009, pp. 25, 51.
3 Ibid., p. 13.
4 Ibid., pp. 7, 9, 13, 17.
5 Plomp n.d.
6 See Dunn and Higgitt 2014.
7 Willms, Kitanov and Langford 2017.
8 See Huygens's entries in Thompson 2001–9.
9 Plomp n.d.
10 Ibid.
11 For Thuret's genealogy, see https://gw.geneanet.org/garric?lang=en&n=thuret&oc=0&p=isaac&type=tree (accessed 20 November 2023).

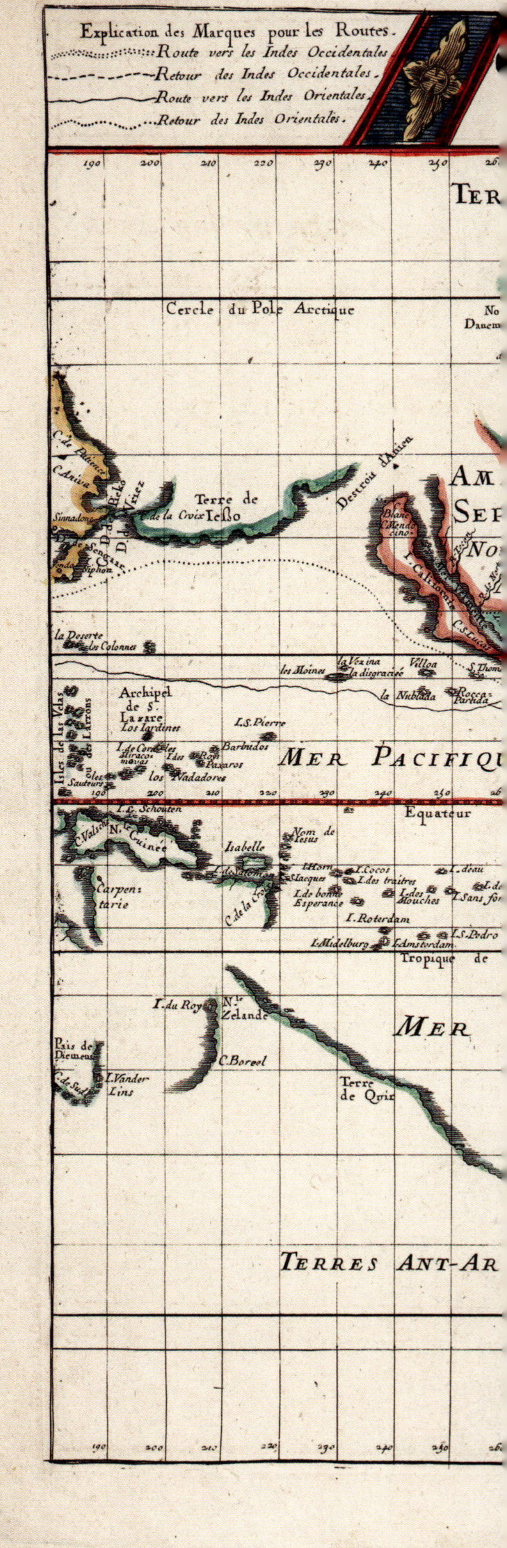

ILL. 15
Carte universelle du commerce ... (Global Map of Trade ...), showing navigation routes for the West and East Indies from France and Spain, by Pierre Du Val, 1686 edition
National Maritime Museum, Greenwich, London. Object no. G201:1/32

Carte Vniverselle du Commerce, c'est à dire
Hidrographique, où sont exactement decrites, Les Costes des 4 Parties du Monde,
Avecque les Routes pour la Navigation des Indes, Occidentales & Orientales.
Par P. Du-Val Geographe Ordinaire du Roy. 1686
A Paris.
Chez l'Autheur, en l'Isle du Palais,
sur le quay de l'Orloge,
au coin de la rüe de Harlay.
Avec Privilege du Roy, pour vingt ans.
Ocean Septemtrional
Mer de Tartarie
Grande Tartarie
Canada ou Nouvelle France
Europe
Asie
Perse
Empire du Mogol
Inde
Chine
Arabie
Barbarie
Afrique
Nubie
Ethiopie
Congo
Mer de Nort
Amerique Meridionale
Bresil
Mer des Indes
Tropique de Capricorne
Nowelle Holande
Ocean Meridional
Mer Magellanique
Australes et Magellaniques et Inconnues
Cercle du Pole Ant-Arctique

3. Cassini's Map of the Moon

RICHARD DUNN

On 18 February 1679, astronomer Jean-Dominique Cassini presented the Royal Academy of Sciences with an impressive printed lunar map (Ill. 16). Although large (lunar diameter 53 centimetres), the map nevertheless did not show the moon as his fellow academicians might have seen it. For one thing, its features appear more crisply than at full moon, when sunlight washes out fine details. For another, its orientation has north at about 5 o'clock and some parts are fancifully imagined. Yet the inclusion of a related drawing behind a globe in Henri Testelin's painting of 1673–81 (see Ill. 9) testifies to how well Cassini's lunar project embodied the Academy's agenda of fostering science in support of the French state.

Born Giovanni Domenico in Italy, in 1650 Cassini became professor of astronomy at the University of Bologna, where he encountered the work of Giuseppe Campani, one of Europe's best lens-makers. Using telescopes by Campani, Cassini soon made discoveries, including Mars's and Jupiter's rotation periods, that brought him fame as a skilled observer.

At this time, Louis XIV's influential minister Jean-Baptiste Colbert was looking to other countries for scholars he might entice to Paris to raise the standard of French science for the nation's benefit. Having encouraged Cassini to become a corresponding member of the Academy, in 1668 Colbert invited him to Paris for a few months. Colbert's invitation, like his courting of Dutch polymath Christiaan Huygens, was motivated by the challenge of determining longitude. Colbert was ready to offer Cassini an even higher salary than Huygens to persuade him to come to France.

For a nation with global ambitions for imperial and commercial expansion, Colbert realised that being able to calculate one's position on land and at sea was crucial. While determining latitude (north–south position) was reasonably straightforward, longitude (east–west position) remained a practical challenge. The principle that one could find longitude by knowing local time simultaneously at two different places was understood. If the sun's position showed that the time was noon for an observer and they could work out that it was also five hours before noon at another specific location, the longitude difference between the two was 75 degrees (since the earth rotates 360 degrees in a day, 15 degrees in an hour). The problem was working out local time anywhere other than where the observer was. This was where astronomy might help, since one could use eclipses or other astronomical events to retrospectively compare times at which the phenomenon was observed from different places, and/or compare the time an event was

ILL. 16
Carte de la Lune de Jean Dominique Cassini (Jean-Dominique Cassini's Map of the Moon), engraved by Jean Patigny after Jean-Dominique Cassini, 1679
Bibliothèque de l'Observatoire de Paris. Object no. Inv.I.1576

ILL. 17
Veüe et perspective de l'Observatoire … (View and Perspective of the Observatory …), engraved by Pierre Aveline the Elder, *c.*1705
Musée national des châteaux de Versailles et de Trianon. Object no. GR 105

observed at one location with tables predicting when it would be seen somewhere else, such as an observatory. Cassini had published tables of the eclipses of Jupiter's satellites for exactly this purpose in 1668, and hence was a recognised longitude expert.

The Italian arrived in Paris in April 1669 and before long his short stay became a lifetime of service, with Cassini becoming a French citizen and changing his name to Jean-Dominique. But it was not until September 1671 that he could move into the Paris Observatory, newly completed as a home for astronomical research by the Academy's members. From then on, Cassini's work included handling large, cumbersome telescopes, often suspended from masts in the Observatory grounds (Ill. 17). An ongoing programme was determining terrestrial longitudes, for which Cassini coordinated the results of expeditions to places such as Denmark, Cayenne (in French Guiana), Egypt, the Cape Verde archipelago and the Antilles, comparing observations made there with his own in Paris.

Cassini began observations for the moon map as soon as he moved into the Observatory. This was another longitude project, intended to be sufficiently detailed to allow observers to compare the times they saw the earth's shadow pass over individual lunar features during an eclipse. Cassini was no great draughtsman and relied on skilled artists – Sébastien Le Clerc and, after a few months, Jean Patigny – to depict the lunar surface based on his observations and notes. The final map, therefore, is a composite of sketches made over eight years, each drawn when a particular place was clearly visible. The aim was not to show the moon as seen at any one time, but rather to render each element precisely.

ILL. 18
Detail of the woman's head, from Cassini's Map of the Moon
Bibliothèque de l'Observatoire de Paris. Object no. Inv.I.1576

Although the map was meant to show the moon faithfully, it includes intriguing flights of fancy, such as a pierced heart and a dragonfly. Most strikingly, the Heraclides promontory, jutting into the Mare Imbrium, is depicted as a woman's head, hair flowing behind as if in a breeze (Ill. 18). But whether the head was really there was debatable. Patigny drew it in at least one of his sketches (Ill. 19). Parisian scholar Henri Justel also mentioned it being observed to German polymath Gottfried Leibniz, while admitting it could be an illusion.[1] Jean de La Fontaine mocked it as a trick of the light in his fable, 'Un Animal dans la Lune' (An Animal in the Moon), first published in 1678.[2] Nevertheless, the head featured in images until the 1890s. Recent research has suggested that it was originally a symbol of love for Geneviève de Laistre, whom Cassini married in 1674.

Although rare in its original version, the map gained some prominence in other editions. Cassini had a simplified version (diameter 16 centimetres) published ahead of an eclipse in 1692. His description exhorted readers to record when

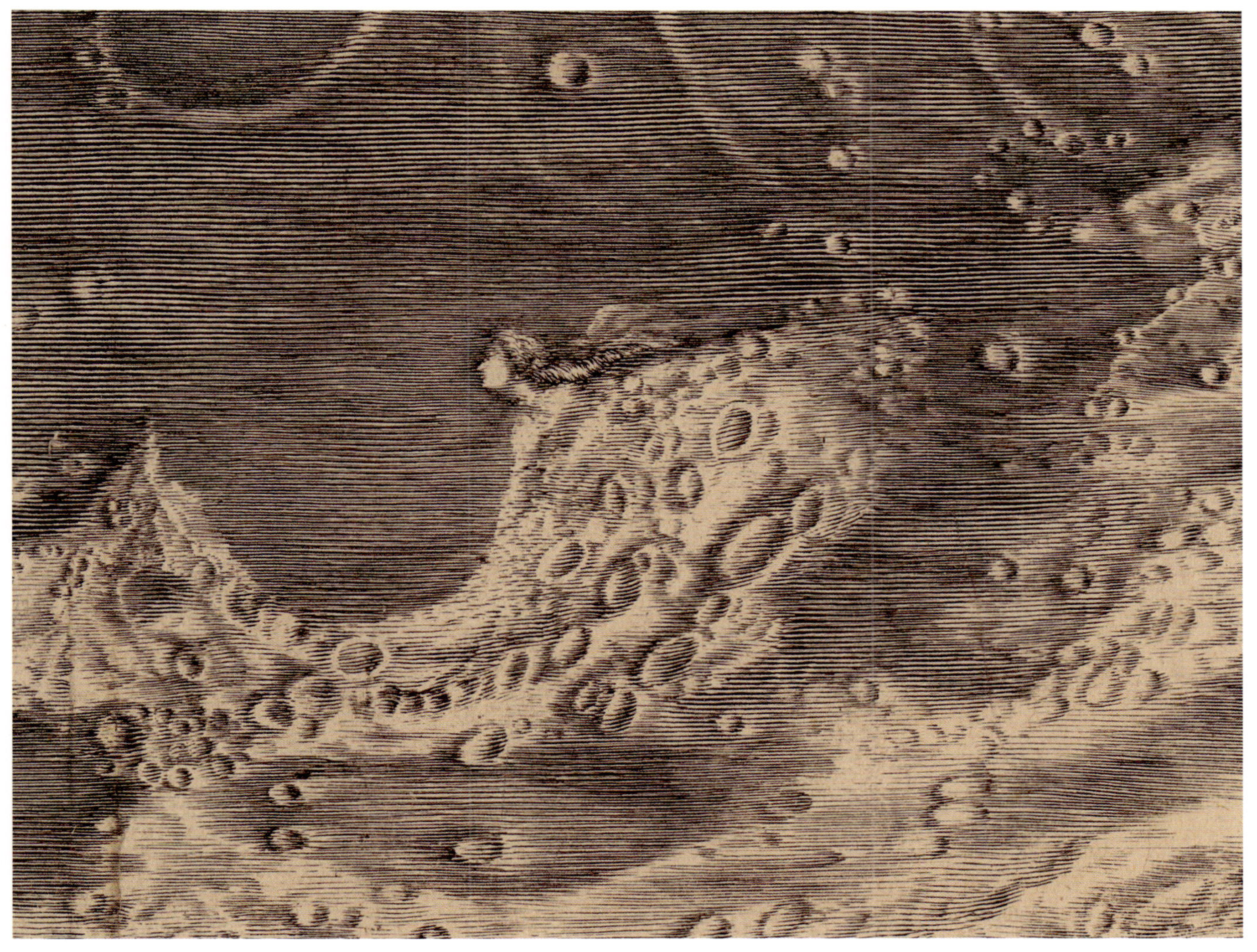

the earth's shadow passed over 40 numbered features in order to determine the longitudes of different observing locations.[3] As he later emphasised, rather than being 'useless descriptions of an imaginary country', lunar maps helped 'perfect geographical and hydrographical maps, without which it is impossible to make long journeys and trade with far-off peoples'.[4] A still smaller version (diameter 13.3 centimetres) appeared regularly in the annually published astronomical tables, *Connoissance des Temps* (Knowledge of the Times), again for determining longitudes. In all the smaller versions the woman's head was no more and the moon's orientation was as if seen through an astronomical telescope (inverted, north at the bottom).[5] In 1787 Cassini's great-grandson, also Jean-Dominique, ordered a new printing from the original copperplate and the following year had a smaller version (diameter 17.6 centimetres) produced with a historical and descriptive commentary. It seems entirely fitting that the fourth of the Cassini dynasty to head up the Observatory chose to reissue and pay homage in this way to one of the most striking pieces of work by his eminent predecessor.

NOTES

1 Henri Justel, letter to Gottfried Leibniz, 13 August 1677, cited in Gislén et al. 2018, pp. 211–25 at p. 215.
2 La Fontaine 1682, Book 2, pp. 32–34.
3 Cassini 1692a, pp. 111–12.
4 Cassini 1692b, pp. 129–35.
5 For example, see Anon. 1750, pp. 196–98 (map after p. 198).

ILL. 19
Preparatory drawing of the Mare Imbrium including the Heraclides promontory, by Jean Patigny, annotated 25 October 1678
Bibliothèque de l'Observatoire de Paris. Object no. D6/40

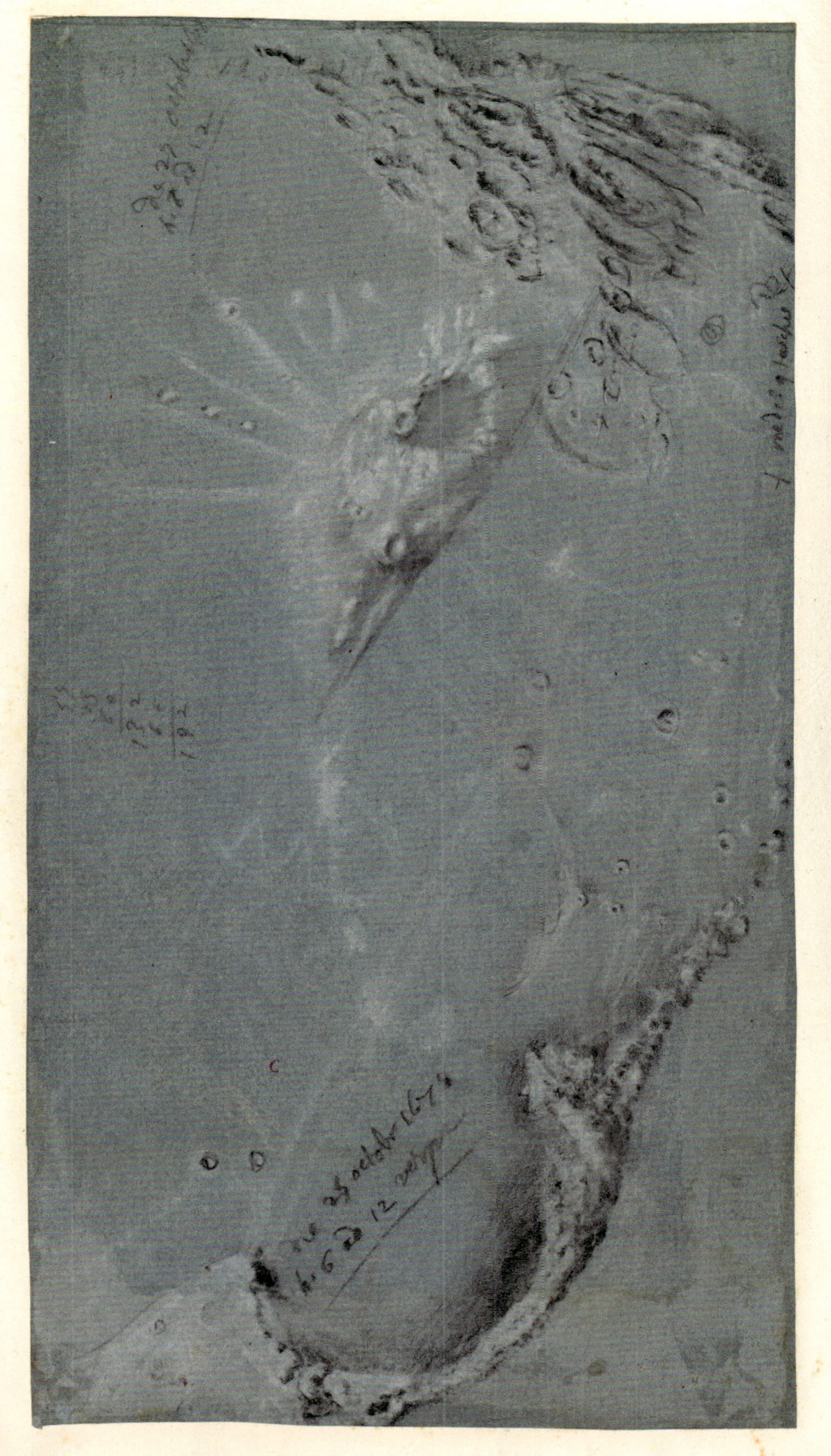

4. Thuret's Machines: A Hymn to the Glory of the King

JULIE GAREL-GRISLIN

The great figures of the eighteenth-century philosophical movement known as the Lumières established France's position on the European intellectual stage. Parisian circles attracted thinkers of all persuasions, who developed new schools of thought influenced by their encounters. Some questioned the role played by private patrons and called for state sponsorship of the arts and sciences.[1] In 1666, on the advice of his minister Jean-Baptiste Colbert, Louis XIV founded the Royal Academy of Sciences and France joined the movement begun by Italy with its Accademia del Cimento in 1657 and England with its Royal Society in 1660. The favourable conditions offered by this new institution encouraged many savants to settle in France, for instance the Dutch physicist Christiaan Huygens to whom we owe the discovery of Titan – the largest of Saturn's moons – and the first exhaustive description of the solar system, and the Italian Giovanni Domenico Cassini, who left the Osservatorio de Panzano (Bologna Observatory) in order to establish one in Paris.

ILL. 20
Planetarium (left) and eclipsarium (right), made by Isaac Thuret after designs by Ole Rømer, 1680–81
Bibliothèque nationale de France, département des Cartes et plans, Paris. Object no. GE A-280 (planetarium); GE A-281 (eclipsarium)

Attracting this papal protégé was no mean feat for Colbert. Cassini was a renowned astronomer, and France spared no expense in order to acquire the skills of such a man. This was a hard blow for its Italian rival: 'in science as in economics and politics the dominant idea of the times was that, in order to secure its wealth in a particular area, a state needed to impoverish its competitors in that field'.[2] Each discovery made by the savants who flocked to Paris contributed to further enhance the king's status and thus heightened the power of France. For Louis XIV, creating an observatory under the leadership of Cassini was not a mere architectural gesture intended to endow the country with a building dedicated to the observation of the heavens; it was essentially about affirming France's strategic and political ambitions by establishing the high importance of astronomical research.[3]

Considered the 'queen of sciences', astronomy held an especially important position.[4] The iconographical programmes of the Palace of Versailles presented an opportunity to glorify the work of the astronomers – for instance, the planned ceiling for the Salon de Saturne in the Grands Appartements was dedicated to the discoveries of Huygens and to Cassini, who discovered two more satellites of Saturn in 1671 and 1672 (Ill. 21). Subtle references to scientific instruments for measuring and calculating were placed or sculpted in the innermost recesses of royal apartments. Astronomy punctuated the daily life of Louis XIV and the members of his court: they never tired of observing the multitude of stellar objects in the heavens.[5] No lunar or solar eclipse appears to have escaped their curiosity. As instruments were

IANVIER
FEBVRIER
MARS
APVRIL
MAI
IVIN
IVILLET
AOVST
SEPTEMBRE
OCTOBRE
NOVEMBRE
DECEMBRE

perfected, observation became more refined and precise. After attentively watching the partial eclipse of 1699 through smoked glass, Louis XIV saw the sun disappear again in 1706, but this time through a giant telescope.

However, astronomy was more than just an activity that appealed to the monarch's own intellectual interests. Thanks to the astronomers, surveys and measurements became more precise. The work of cartographers who relied on them improved in accuracy.[6] The progress in these two areas was of immediate use for Louis XIV's plans: territory that was known and mapped could be administered, exploited and controlled. The map of the 'Paris surroundings' drawn using the geodesic triangulation method perfected by Jean Picard allowed the implementation of Colbert's fiscal and cadastral policies. Its remarkable accuracy made it the prototype for the Cassini map of France ordered from the Academy by Louis XIV in 1671, a colossal enterprise that continued through three reigns (Ill. 22).[7]

ILL. 21
Le Triomphe de Saturne (The Triumph of Saturn): design for the ceiling in the Salon de Saturne at Versailles, by Noël Coypel, *c.*1671–72
Musée national des châteaux de Versailles et de Trianon. Object no. MV 8929

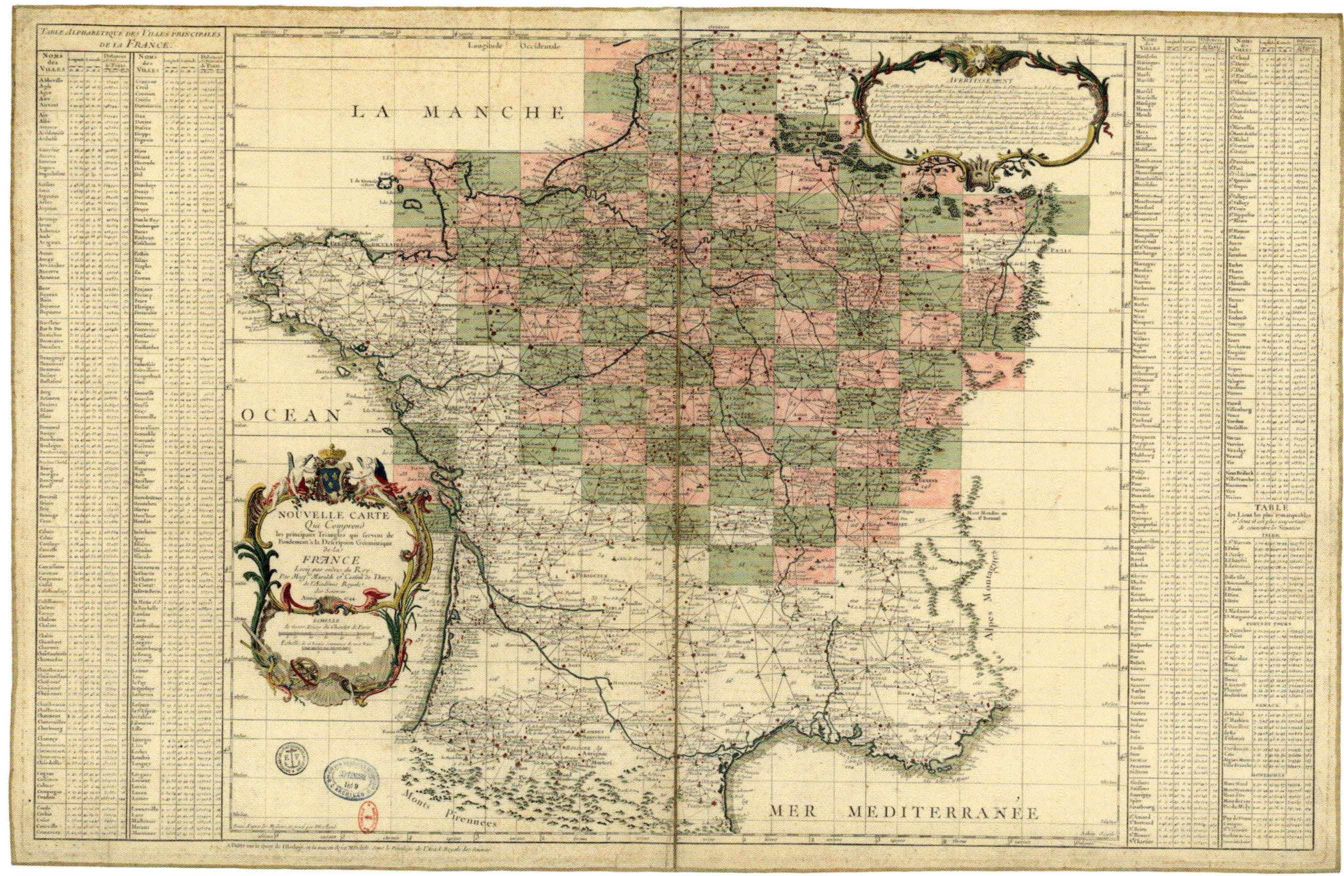

ILL. 22
Nouvelle carte ... de la France (New Map ... of France), engraved by Guillaume Dheulland after Jean-Dominique Maraldi and César-François Cassini de Thury, with lettering by Aubin, 1744
Bibliothèque nationale de France, département des Cartes et plans, Paris. Object no. GE SH 18 PF 1 QUATER DIV 10 P 0

At sea, navigational instruments made it possible to cross the oceans and mark the coasts accurately; blank zones on the maps were filled in with information and new routes were opened up. If mastery of the seas was a considerable advantage for the great European powers, maps were an indispensable aid for a strong navy. The quest for accurately determining longitude at sea was a key subject of debate throughout the eighteenth century and underlines the high stakes of precision mapping.[8]

Sciences, especially astronomy, played an important role in royal education. Louis le Dauphin, known as le Grand Dauphin – the heir apparent of Louis XIV – was surrounded by the greatest scientific minds, and Ole Rømer, a young Danish savant who had recently succeeded in calculating the speed of light, was charged with teaching him the principal ideas in this field. In May 1680, in a letter he sent to the English philosopher John Locke, the intellectual Nicolas Toinard mentioned one of the educational tools that Rømer had developed for introducing Louis to the Copernican system.[9] A few weeks later, in August, Rømer informed his peers at the Academy about his work; he described a 'machine for the planets' and a 'machine for eclipses' that could determine the dates of these astral phenomena over two centuries (Ill. 20). The making of these two 'machines' was entrusted to a famous artisan, Isaac Thuret, the royal clockmaker, who was responsible for maintaining the Academy's instruments. The resulting creations were astounding and the fact that they were much admired by James II of England when he visited Paris is significant if we consider this scientific, technical and aesthetic achievement in the context of European competition. The splendour of these objects and their symbolic potential did not escape the Jesuits whose task it was to build scientific, diplomatic and commercial relationships in Asia, particularly with the Chinese Empire. They were well acquainted with the importance of astronomy in China, where it was a true 'state' science.[10] In the 1670s the Belgian Jesuit missionary Ferdinand Verbiest, who was received at the Beijing court, produced six bronze instruments to be placed on the terrace of the Peking (Beijing) Observatory, which had been founded during the Ming dynasty (1368–1644) (Ill. 23). It seems likely that copies of the planetarium and eclipsarium designed by Rømer were made and presented as diplomatic gifts to the Kangxi emperor.[11]

The three dolphins supporting the feet of the two machines quite clearly recall the initial purpose of the project, the education of the crown prince, whose title of Dauphin also means 'dolphin' in French. The ebony and gilded bronze forms highlight the instruments' fine details, and it is easy to recognise the influence of the Boulle school of cabinetmaking.[12] Designed as a hymn to the glory of Louis XIV, the machines are adorned with symbols exalting his grandeur. The globe, lyre, horns of plenty and scientific instruments refer to Apollo, the protector of the arts and god of the sun. Atop both machines, these emblems are associated with Louis XIV's radiant and crowned Sun King motif, which evokes at the same time the heliocentric system of Copernicus, the Sun King and his almighty power.

NOTES

1 Hahn 2010–11, pp. 31–38.
2 Ancelin 2011, p. 37.
3 Wolf 1902.
4 Delalex 2022, pp. 163–69.
5 Widemann 2010–11, pp. 220–24.
6 Grimbergen 2004, pp. 91–102; Major 2014, pp. 113–29.
7 Pelletier 1986, pp. 26–32.
8 Despoix 2000, pp. 205–33; Mahoney 1980, pp. 234–70; Jullien 2002.
9 Turner 2018, pp. 114–22.
10 Jami 2017, pp. 65–74.
11 Castelluccio 2019, pp. 25–44.
12 Sarrazin 2003, pp. 46–47.

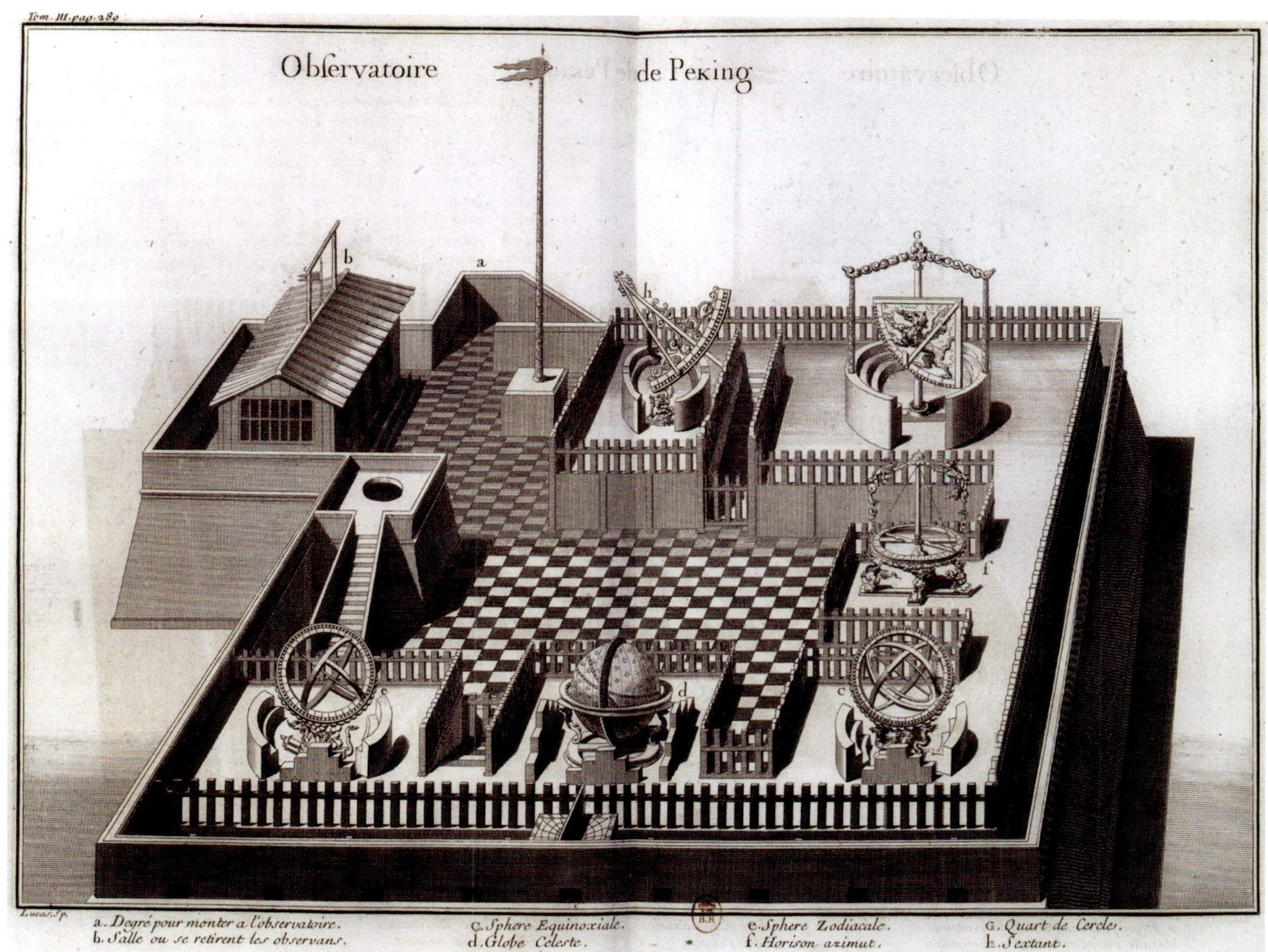

ILL. 23
View of the Peking (Beijing) Observatory, a plate from *Description … de l'Empire de la Chine et de la Tartarie chinoise …* (A Description of the Empire of China and Chinese-Tartary …), by Jean-Baptiste Du Halde (Paris, 1735), vol. 3
Bibliothèque nationale de France, département des Estampes et de la photographie, Paris. Object no. OE-1B-PET FOL

5. The Marly Machine: The Quest to Bring Water to Versailles

MATTHEW HOWLES

'I see the Seine suspended aloft', proclaimed a 1697 ode dedicated to Louis XIV.[1] This evocative image of a river impossibly flowing through the air praised one of the most ambitious hydraulic engineering projects of his reign: the Marly Machine.

The brainchild of two men from Liège – entrepreneur Arnold de Ville and carpenter-engineer Rennequin Sualem – the Marly Machine was designed to supply water from the River Seine for the gardens of the royal palaces of Marly and particularly Versailles. Built between 1681 and 1684, its origins are intimately connected with the official establishment of Versailles as the centre of Louis XIV's court in 1682. A vast construction programme transformed what had been a relatively modest hunting lodge for his father and predecessor, Louis XIII, into a suitably magnificent statement of royal power: the impressive palace known today. This work began with the grounds, which from 1661 were engineered into sweeping vistas, stepped terraces and geometrically ordered layouts demonstrating the king's authority to control nature itself. For Louis XIV, the scale, splendour and spectacle of Versailles's gardens also provided an opportunity to rival and transcend prevailing European fashions for Italian garden design – an assertion of French political pre-eminence in physical form.[2] Key to his vision was the creation of fountains and waterworks unsurpassed in their aesthetic flamboyance and technological sophistication (Ill. 25).

Versailles has indeed been famed for its water features since Louis XIV's reign – especially remarkable given the absence of a suitable, nearby water source. Many schemes of colossal scale and ambition sought to redress the issue. In 1674 the astronomer abbé Jean Picard of the Royal Academy of Sciences was tasked with surveying and levelling the local topography, drawing water downhill from rivers, streams and lakes into an extensive network of vast new reservoirs to feed Versailles below. Applying his experience and expertise in astronomy, Picard designed a surveying level fitted with a telescope that could take sufficiently accurate angle and elevation measurements over the significant distances involved (Ill. 26). The most extraordinary hydraulic scheme of all, though, was the Marly Machine, the largest mechanical device of its day.[3]

While Picard's levelling works relied on gravity, calculating downhill routes for water to reach Versailles from higher sources, the Marly Machine had to defy it by raising water from the Seine, which lies below the palace. The Machine itself no longer exists, but it has been immortalised through surviving documents and material culture from the period, including prints, drawings, maquettes, medals,

ILL. 24
Vue de la machine de Marly et du château de Louveciennes (View of the Marly Machine and the Palace of Louveciennes), by Pierre-Denis Martin, 1722–23
Musée national des châteaux de Versailles et de Trianon. Object no. MV 778

eyewitness accounts, panegyric odes, engineering manuals and the famous *Encyclopédie*.[4] It also lives on in a 1722–23 oil painting, one of eight views of royal residences commissioned by Louis XV from the artist Pierre-Denis Martin to decorate apartments at Versailles (Ill. 24).

Martin's painting shows the Machine looming from across the Seine, where a chain of natural islands had been artificially reinforced into a continuous belt under the direction of Louis XIV's military engineer, Sébastien Le Prestre de Vauban. Thus divided and with its power concentrated, the river accommodated fishing and boat passage via one channel and the Marly Machine via another. Entering the Machine from upstream (to the left of the painting), the Seine's flow turned 14 enormous paddle wheels each about 12 metres in diameter. The wheels then drove a crank and connecting rod system, shown snaking its way up the hill behind the Machine in two parallel chains. This mechanism powered roughly 260 pumps that forced the water upwards through cast-iron pipes in three stages, over an impressive total elevation gain of 162 metres: from the Seine to a reservoir at the hill summit via an intermediary reservoir halfway up, and from there higher still to the Louveciennes aqueduct. Stretching off into the distance of the painting, the aqueduct carried the water as it began its downhill journey to Versailles some

ILL. 25
Vue perspective du château de Versailles depuis le bassin de Neptune (Perspective View of the Palace of Versailles from the Bassin de Neptune), by Jean-Baptiste Martin the Elder, 1690–1700
Musée national des châteaux de Versailles et de Trianon. Object no. MV 751

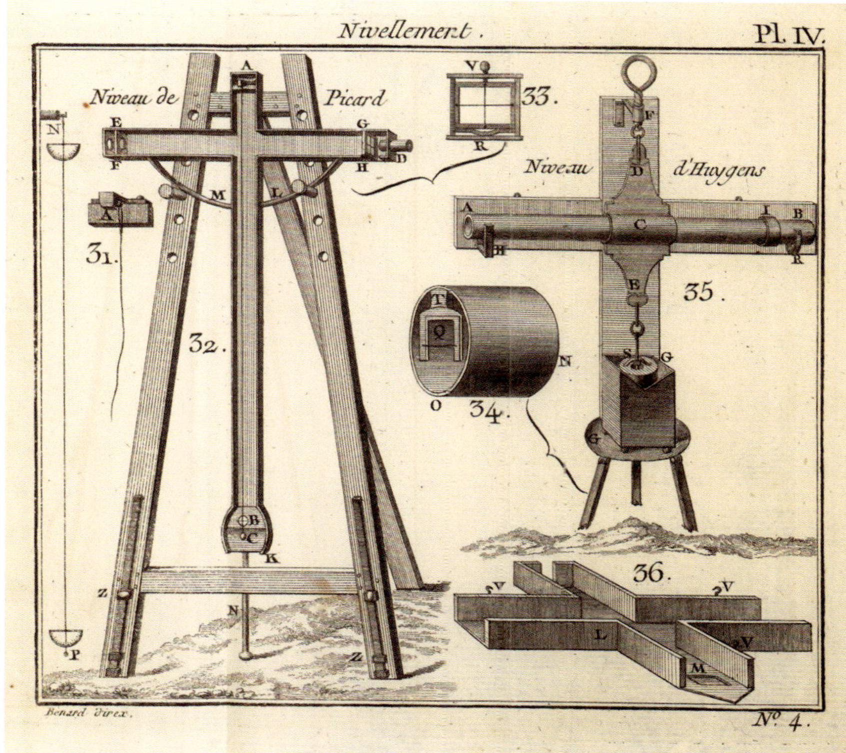

ILL. 26
Abbé Jean Picard's surveying level comprising a telescope fitted with sights mounted on a tripod with plumb-bob (Fig. 32, 'Niveau de Picard'), a plate from his book *Traité du nivellement* (Treatise on Levelling) (Paris, 1780 edition)
Science Museum Group, London. Object no. 9900113647

7 kilometres away. Buildings surrounding the Machine on the riverbank and hillside housed forges, carpentry workshops, warehouses and accommodation, testament to the substantial supporting infrastructure and workforce needed to maintain it. Martin's painting captures the sheer monumentality of the Machine, populated with figures dwarfed by its hulking wooden structure as water cascades from the blades of its rotating wheels. Up to 1,800 men were involved in its construction, the cost of which rose to almost 4 million livres – about £105 million today.[5]

The Marly Machine and other hydraulic schemes provided a store of more than 8 million cubic metres of water to feed Versailles. Ultimately, though, they were unable to satisfy Louis XIV's demands. Not enough water could be supplied and replenished at sufficient speed to keep the fountains playing constantly. Fountaineers were instructed to turn the water features on and off as king and company passed, to at least give the illusion of continuous spectacle. These initiatives nevertheless represented the immensely ambitious feats of science, technology and engineering that were driven in Louis's name. Indeed, the Marly Machine was as much an expression of Louis XIV's wealth and power as the fountains it sustained. Louis understood that creating the most impressive gardens and waterworks in Europe would transform Versailles into a visible representation – and physical manifestation – of his power.[6] He even designed itineraries or *promenades* to direct important visitors around the gardens, to showcase their grandest elements and by extension his own glory and magnificence. A 1687 engraving depicts him receiving three ambassadors from Siam (Thailand) at Versailles in September 1686, his authority clear: he

L'AVDIENCE DONNÉE PAR LE ROY AVX AMBASSADEURS DU ROY DE SIAM A VERSAILLES LE PR.ER SEPTEMBRE . 1686.
LES TRIOMPHES DE L'EGLISE SOUS LE REGNE DE LOUIS LE GRAND,
et les Conquestes des Chrestiens sur les Infideles en l'année 1686.
BUDE
MANDARINS
le Danube Fl.
ALMANACH POUR L'ANNÉE M. DC. LXXXVII.
A PARIS, chez NICOLAS L'ANGLOIS, rüe S. Jacques, à la Victoire. Avec Privilege du Roy.

ILL. 27
L'Audience donnée par le Roy aux ambassadeurs du Roy de Siam à Versailles le Pr[r] septembre 1686 (Audience Given by the King to the Ambassadors of the King of Siam (Thailand) at Versailles on 1 September 1686), published by Nicolas Langlois the Elder, 1687
Bibliothèque nationale de France, département des Estampes et de la photographie, Paris. Object no. RESERVE QB-201 (171)-FT 5 [Hennin, 5551]

sits elevated as they gesture reverentially below, while other members of their entourage bow to the ground (Ill. 27). Significantly, the backdrop opens on to the gardens of Versailles and their resplendent water features, of which the Siamese party were given an extensive tour. They even visited the Marly Machine itself, when one of the ambassadors is reported to have asked, 'Is it a man or a demon who has built this machine?' He concluded: 'This work is due to the greatness of the king'.[7]

NOTES

1 'Interim pendet per inane ductus Sequana | J'apperçois dans les airs la Seine suspenduë'. Boutard 1697. English translation by the present author.

2 Saule and Ward 2016–17, p. 165; Morera 2010–11, p. 87. On Versailles's gardens as manifold expressions of royal and political power, see especially Mukerji 1997.

3 Brandstetter 2012, para 1. An even bolder project sought to divert the River Eure towards Versailles, but was never completed. See Sabatier 2017, chapter 7, paras 31–35.

4 Diderot 1762–72, plates vol. 5 (1767). On other surviving material, especially written sources, see Brandstetter 2012.

5 Particulars and statistics about the Machine are taken from Etablissement public du château, du musée et du domaine national de Versailles, secteur éducatif n.d., pp. 3–20.

6 On this subject, see particularly Brandstetter 2012 and Mukerji 1997.

7 'Est-ce un Homme, ou un Demon qui a fait cette Machine? … Cét Ouvrage est dû à la grandeur du Roy'. See Donneau de Visé 1686, pp. 267–68. English translation by the present author.

6. Between Science and Diplomacy: Cultivating Pineapples at Versailles

HÉLÈNE DELALEX

ILL. 28
Ananas dans un pot (Pineapple in a Pot), by Jean-Baptiste Oudry, 1733
Musée national des châteaux de Versailles et de Trianon. Object no. MV 7035

Framed by a dark stone niche, this pineapple in a pot is displayed in regal splendour (Ill. 28). The fruit is the sole subject of this canvas painted in 1733 by Jean-Baptiste Oudry to celebrate the growing of the first pineapple in the King's Kitchen Garden at Versailles that year. Originating from the New World, chiefly South America and the Caribbean, the pineapple was discovered during voyages of exploration in the late fifteenth century, then described and brought to Europe in the early sixteenth century. Gonzalo Fernández de Oviedo y Valdés, an envoy of the King of Spain, produced the first description of it in 1535, accompanied by an illustration.[1] Forty years later the French traveller Jean de Léry brought pineapples back from Brazil and gave this rapturous description:

> They are bluish yellow in colour and give off such a strong scent of raspberry that, when walking in the woods and other places where they grow, you can smell them from a very long way off. What is more, they melt in the mouth and are naturally so sweet that no preserves in France can surpass them.[2]

The first pineapples brought to Europe by explorers were barely able to withstand the hazards and the length of the journey. However, people were fascinated by them as can be seen in a famous painting in the British Royal Collection, which shows the royal gardener John Rose on his knees presenting King Charles II of England with a pineapple that had survived a long journey (Ill. 29). The sovereigns of Europe wanted to own this prestigious fruit in their royal gardens, the more so because it was extremely difficult to acclimatise and very sensitive to the cold and damp – growing it successfully was considered a scientific achievement at the cutting edge of horticulture. Thanks to their innovative glasshouses, the Dutch were the first to succeed in bringing pineapples to the fruiting stage in the late seventeenth century, while in Sweden a pineapple from India reached maturity on 30 July 1729 at the royal Ulriksdal Palace, north of Stockholm (Ill. 30).

It was not until four years later that France achieved this much-desired botanical success. After several fruitless attempts, in the winter of 1733 Louis Le Normand, director of the King's Kitchen Garden, managed to grow two pineapples from American cuttings presented to the king. During the reign of Louis XV, the expansion of the French colonial empire in South America, which caused new rivalry with England, favoured expeditions to and exports from the Antilles. Because of the length of time a pineapple plant requires to reach maturity

J.B, oudry
1733
567

and produce fruit, Le Normand had a Dutch hothouse installed where the potted plants were placed in warm layers of oak bark and arranged facing the light that entered through large panes of glass equipped with blinds of green waxed cloth. These were essential for controlling the amount of light and collecting the condensation to maintain a dry climate. The ambient temperature was never allowed to drop below the limit of 12°C, and after October these greenhouses were heated by woodburning stoves. Thus, on Christmas Day, Le Normand proudly presented the king with the first pineapples from Versailles.[3] The sovereign admired their beauty before praising their scent and their exquisite flavour.

ILL. 29
Charles II Presented with a Pineapple, c.1675–80
Royal Collection Trust, London. Object no. RCIN 406896

Louis XV was captivated and wished to develop large-scale cultivation of this plant in the gardens of the royal palace of Choisy, not far from Versailles. In 1749–50 a new grand kitchen garden of 15 arpents (75 hectares) was laid out for

ILL. 30
Ananas i kruka (Pineapple in a Pot at Ulriksdal Palace), by David von Cöln, 1729
Nationalmuseum Stockholm. Object no. NMGrh 1308

this purpose.[4] The first pineapple houses were built in 1754, and five years later the marquis de Marigny, the director of the Bâtiments du roi (the King's Buildings), organised the cultivation of 400 pineapples.[5] In order to ensure the success of this venture, in the summer of 1763 the king called on an English specialist, the gardener Alexandre Brown, who had placed his talents at the service of the duc de Belle-Isle, a Marshal of France. In the year of his arrival he harvested his first pineapple, which he immediately sent to the king who was staying at Compiègne.[6] Brown, certain of his talent and enjoying the king's trust, was a key figure in expanding the cultivation of the pineapple, but at extraordinary and ever-increasing costs. Shortly afterwards, in 1765, Claude Richard, director of the botanical garden at Trianon, perfected a new pineapple hothouse, which was installed in this garden.[7]

ILL. 31
The Grand Ananas textile: modern recreation by Maison Pierre Frey after the original produced at the Manufacture Royale de Jouy in 1784 and owned by the Musée de la Toile de Jouy, used to redecorate Marie-Antoinette's private chambers at Versailles in 2023

The pineapple sparked an extraordinary fashion at the courts of Europe and among rich amateurs, and as had happened with chocolate in the preceding century, the newly popular fruit was credited with every therapeutic virtue. Moreover, with its golden colour and crown of leaves, the pineapple was considered to be the 'king of fruits'. 'This is probably the reason why the King of Kings placed a crown on its head, an essential symbol of its royalty', wrote the Reverend Father Du Tertre.[8] In the context of a general infatuation with exoticism, it naturally became a source of inspiration for artists: its perfect oval form, the lozenge shapes on its skin and the exuberance of its uppermost leaves presented the decorative arts with a new motif. Pineapples appeared as decoration on the leaves of fans and the textiles covering furniture and hung in apartments, such as the fabric known as the Grand Ananas (the great pineapple) produced by the Manufacture Royale de Jouy (Royal Manufacture of Jouy). This textile, which now decorates the second floor of the private chambers of Marie-Antoinette, echoes the fine portrait of a pineapple chosen by the queen herself to decorate the overdoor of her *grand cabinet intérieur* at Versailles (Ill. 31).

NOTES

1 Fernández de Oviedo y Valdés 1535, part 1.
2 Léry 1578, p. 211.
3 Le Roi 1847, p. 13.
4 Rouet 2019, section 7.
5 Lamy 2011, p. 108.
6 National Archives, Paris, O1 1347/226, quoted by Rouet 2019, section 46.
7 *Profil pour une serre d'ananas suivant le nouveau système préconisé par le sieur Richard, avec l'indication des ombres portées au solstice d'hiver et au solstice d'été* (Profile of a Pineapple House Following the New System Recommended by Lord Richard, Showing the Shadows Provided at the Winter Solstice and the Summer Solstice), 1765, pen and brown ink, National Archives, Paris, CP O1/18871, no. 82.
8 Du Tertre 1667–71, vol. 2 (1667), p. 127.

7. Louis XV's Rhinoceros

JOSÉPHINE LESUR

This specimen of a male Indian rhinoceros (*Rhinoceros unicornis*) was presented to Louis XV in 1769 by Jean-Baptiste Chevalier de Conan, the governor of Chandernagore (now Chandannagar) in Bengal (Ill. 32). Following the creation of the French East India Company in 1664 by Louis XIV's minister Jean-Baptiste Colbert, France installed various trading posts in India, including one in Chandernagore in 1686. The purpose of these establishments, the largest of which was in Pondicherry (now Puducherry) on the eastern coast, was to facilitate the supply of Asian products such as silk and pepper and to limit the influence of the English and the Dutch, who were already long-established in the region. Although the Seven Years War (1756–63) considerably reduced the French possessions in India, Chevalier, who held the post of governor between 1767 and 1778, nevertheless tried to consolidate the French positions there. When he sent the rhinoceros to France in 1769, he had been governor of Chandernagore for two years and had a good knowledge of the region where the rhinoceros originated, in the foothills of the Himalayas. In the 1750s he journeyed into the north-east of India, where he encountered wild animals such as tigers, elephants and rhinoceros that made a deep impression on him.

ILL. 32
Indian rhinoceros presented to Louis XV in 1769, stuffed in 1793
Muséum national d'Histoire naturelle, Paris. Object no. MNHN-ZM-MO-1991-1439

In addition to their roles of administrator and representative of the king, governors were expected to provide exotic plants and animals to stock the menagerie of Versailles and the King's Garden in Paris (now the Jardin des Plantes). The royal menagerie was one of Louis XIV's major projects at Versailles at the same time as the enlargement of the palace. It was built between 1662 and 1664, at the south end of the future Grand Canal along the road between Versailles and Saint-Cyr, but would not be completely finished until 1668. It was centred around an octagonal pavilion referred to as a *château* and also included a gallery surrounded by seven courtyards fitted out to house the animals (Ill. 33). Conceived as a showpiece, its purpose was to demonstrate the monarch's ability to send for and collect zoological rarities from all over the world. Besides being a place of splendour and wonder, the menagerie had a political dimension as a site where the king displayed his power to official visitors from across Europe.

In order to do so, Louis XIV charged Colbert with sending him rare and exotic animals. Colbert organised an entire purchase and transport network via the governors of the French provinces and the French East India Company. The menagerie also housed the animals presented to the king as diplomatic gifts, such as a cow elephant from the Congo, given in 1668 by the regent of Portugal, later King Peter II, and big cats presented by Arab princes.

Consequently, during the period of its prosperity, thousands of animals from Europe, Asia, Africa and America lived in the menagerie. They were a source of inspiration for numerous artists, such as the Flemish animal painter Pieter Boel, whose studies included a painting of one of the rarest and most admired birds in the menagerie, the cassowary, sent to Versailles in 1671 by the French governor of Madagascar, François de Lopis, marquis de Mondevergue (Ill. 34).[1] The animals were also specimens to be studied by many scientists and the menagerie played a part in the vast project of the early years of the Royal Academy of Sciences that had been established in 1666. The enormous variety of European and non-European species provided an inexhaustible supply of specimens for dissection by the anatomists and enabled significant advances to be made in the knowledge of organisms and the characteristics of different animals. The development of techniques for spirit and osteological preparations, as well as taxidermy, made it possible to preserve specimens, some of which were kept in the King's Garden.

However, by the time Chevalier sent the rhinoceros to Versailles, the menagerie had lost some of its splendour. Because Louis XV was not greatly interested and

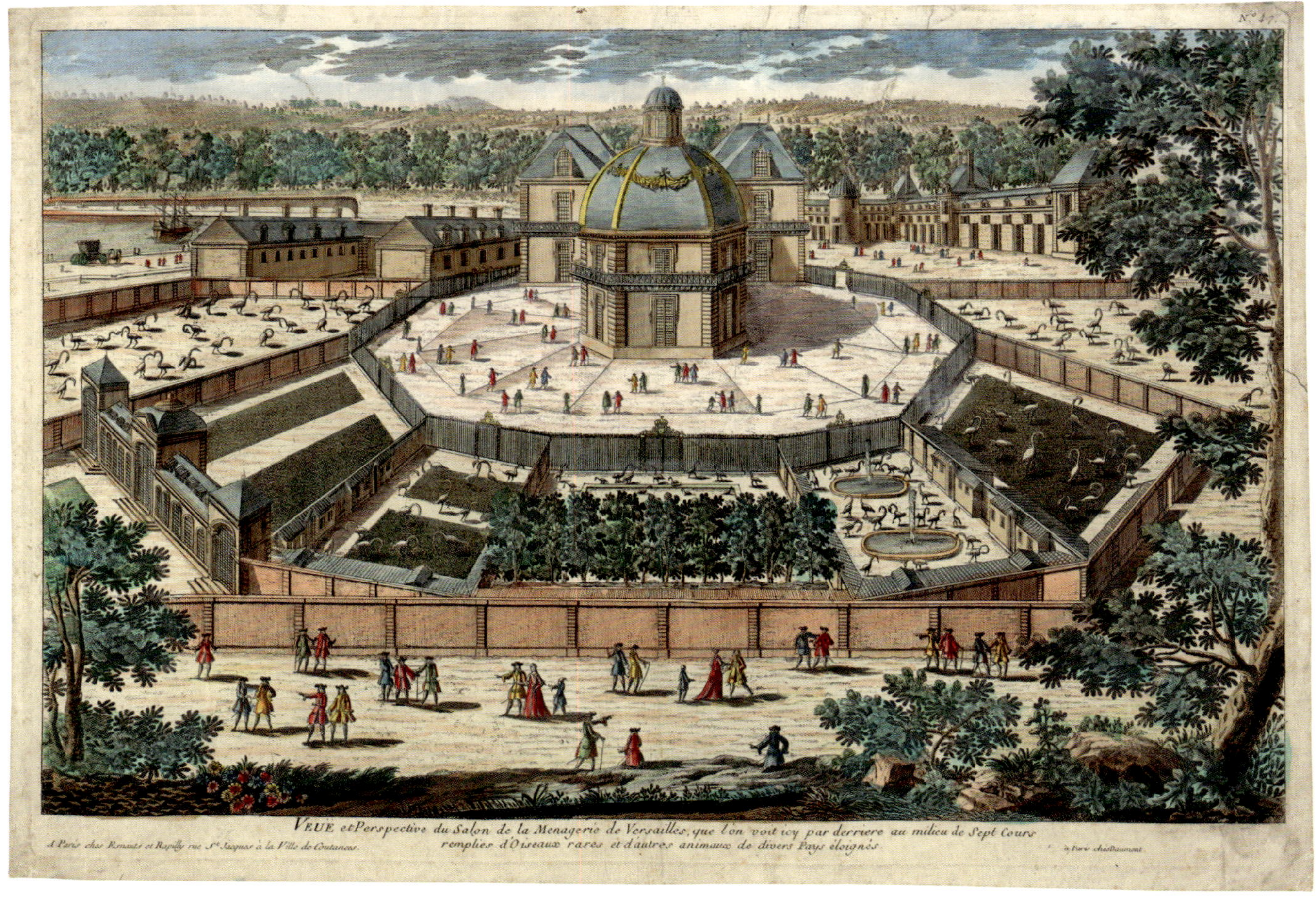

ILL. 33
Veuë et perspective du salon de la Menagerie de Versailles … (View and Perspective of the Pavilion of the Versailles Menagerie …), engraved by Pierre Aveline the Elder, 1770–75 edition
Musée national des châteaux de Versailles et de Trianon. Object no. INV.GRAV 4147

also as a result of the conflicts in Europe, the supply of exotic animals became more limited during the eighteenth century. The rhinoceros arrived after a long voyage. It set off from Calcutta (Kolkata) on 22 December 1769 on the *Duc-de-Praslin*, a ship belonging to the French East India Company, landed at Lorient in Brittany on 11 June 1770, and eventually reached the Versailles menagerie on 11 September 1770.

On its arrival, the rhinoceros was installed in a specially built enclosure, about 20 metres long and 12 metres wide and furnished with a pool. Like its fellow creature Clara, the female Indian rhinoceros that arrived in 1741 and was paraded around Europe by a Dutch sea captain as a fairground exhibit, the Versailles rhinoceros quickly became a popular attraction.[2] Its enormous size, strange appearance and rarity delighted and aroused the curiosity of the aristocracy and the crowds of spectators who travelled from all over the country to see it.

It also captured the attention of many naturalists, such as Georges-Louis Leclerc, comte de Buffon, who came to study it on several occasions. Buffon was an eminent naturalist who administered the King's Garden from 1739 to 1788. He increased its size by more than a third, sent for plants from all over the world, and expanded the king's Cabinet of Natural History, which he would later make into one of the greatest collections in Europe. At the same time he wrote

ILL. 34
Study of the cassowary sent to Versailles, with a white crow or raven, by Pieter Boel, 1671–74
Musée du Louvre, Paris, département des Peintures. Object no. INV 3972

numerous works, including his famous *Histoire naturelle*, which runs to 36 volumes. The king's rhinoceros was rather aggressive. Later, in 1804, Georges Cuvier, Professor of Comparative Anatomy at the Muséum national d'Histoire naturelle (National Museum of Natural History), reported that 'It killed two young men who had unwisely entered its enclosure'.[3]

In 1792, three years after the start of the Revolution, the Versailles menagerie housed only a few animals: a lion from Senegal, a quagga, a hartebeest ... and the old rhinoceros. Louis-Charles Couturier, the supervisor general of Versailles, invited Jacques-Henri Bernardin de Saint-Pierre, the new director of the Jardin des Plantes in Paris, to rescue the animals that were still living at Versailles. The latter proposed to the National Assembly that it should vote for the creation of a menagerie in the Jardin des Plantes. However, all this took time and the rhinoceros died before being transferred to Paris. It was in fact found dead on 23 September 1793, drowned in its pool or, according to some sources, killed by a sword blow.

Its remains were transferred to Paris, to the National Museum of Natural History, founded in 1793. Jean-Claude Mertrud, Professor of Animal Anatomy, assisted by the anatomist Félix Vicq d'Azyr and the naturalist and first director of the museum, Louis-Jean-Marie Daubenton, carried out the dissection. In 1793 the stuffing of an animal of this size was a first in taxidermy and the anatomical result may appear strange: for instance, the way the legs are positioned makes them look like table legs and the rib cage is too round because it is supported by wooden hoops. At the same time as the skin was stuffed, the animal's skeleton was assembled by Mertrud, and was later exhibited in the Galerie d'Anatomie Comparée (Gallery of Comparative Anatomy) when it was established at the end of the nineteenth century, where it can still be seen today (Ill. 35).

ILL. 35
The skeleton of Louis XV's Indian rhinoceros
Muséum national d'Histoire naturelle, Paris. Object no. A7974 (= BVI-63)

NOTES

1 Guerrini 2015, p. 177.
2 Clara was offered to Louis XV, who tried to acquire her but had to give up the idea because of the high price demanded.
3 Rookmaaker 1983, p. 311.

8. Pierre-Joseph Buc'hoz and the Trianon Garden

GABRIELA LAMY

In Haarlem in 1780, a new and exceptionally beautiful hyacinth variety, a flower originating from the Middle East, was bred. As it was customary to name the most beautiful flowers in honour of princes and princesses, the Dutch florists named it 'Maria Antonia Reine de France' (Marie-Antoinette, Queen of France). All flower lovers were impressed by the new hyacinth's 20 double florets set in a pyramid form, with white petals and a full, purple heart. Pierre-Joseph Buc'hoz, a physician and botanist in the employ of Monsieur, brother of Louis XVI, published this engraving of the plant in 1781 in *Collection coloriée des plus belles variétés de jacinthes qu'on montre aux curieux dans les jardins fleuristes d'Harlem* (Coloured Collection of the Most Beautiful Hyacinth Varieties on Show to the Curious in the Flower Gardens of Haarlem) and then in 1783 in *Le Jardin d'Eden, ou le paradis terrestre renouvellé dans le jardin de la Reine à Trianon* (The Garden of Eden, or the Earthly Paradise Recreated in the Queen's Garden at Trianon) (Ill. 36).

ILL. 36
The hyacinth named after Marie-Antoinette, a plate from *Le Jardin d'Eden, ou le paradis terrestre renouvellé dans le jardin de la Reine à Trianon* ... (The Garden of Eden, or the Earthly Paradise Recreated in the Queen's Garden at Trianon ...), by Pierre-Joseph Buc'hoz (Paris, 1783)
RHS Lindley Collections, London. Object no. 999 (4F) TRI BUC, 3647-1001

The *Garden of Eden* is a collection of 200 plates depicting plants without commentary. It illustrates the acclimatisation of the rare plants that were cultivated at Trianon, most of which required hothouses. The plants originated from Guiana and the Antilles, Africa, the East Indies, China and Japan, with some from North America.[1] After 1749, botanists at Trianon began exploring the new economic exploitation of plants from both metropolitan France and the French colonies; the gardens at Trianon were created by Louis XIV, enlarged by Louis XV and finally transformed by Queen Marie-Antoinette (Ill. 37). Success with bringing the pineapple to the fruiting stage in 1733 was the catalyst for building the first hothouses at Versailles, allowing exotic plants to be grown and propagated. In the King's Kitchen Garden at Versailles, on a terrace close to the fig orchard, Philibert Orry, the new director of the King's Buildings, ordered the construction of a hothouse for pineapples in 1737, followed by three more in 1751 for bananas, coffee and palms.

At that time the energy used to heat the hothouses came from the sun (via a framed glass roof aligned to the winter and summer solstices), the burning of wood (through hot-air ducts from stoves and ovens) and the fermentation of biomass (dung heaps). At Trianon the first Dutch hothouse was built in 1749 and depicted in 1762 in the *Encyclopédie*.[2] Claude Richard, a merchant florist from Saint-Germain-en-Laye, was charged with the task of maintaining it. He became the king's gardener at Trianon and enjoyed special status, receiving orders only from Louis XV himself and being paid from the king's personal budget. A true research laboratory – for the benefit of horticulture and botany – was established

MARIA ANTONIA REINE DE FRANCE.

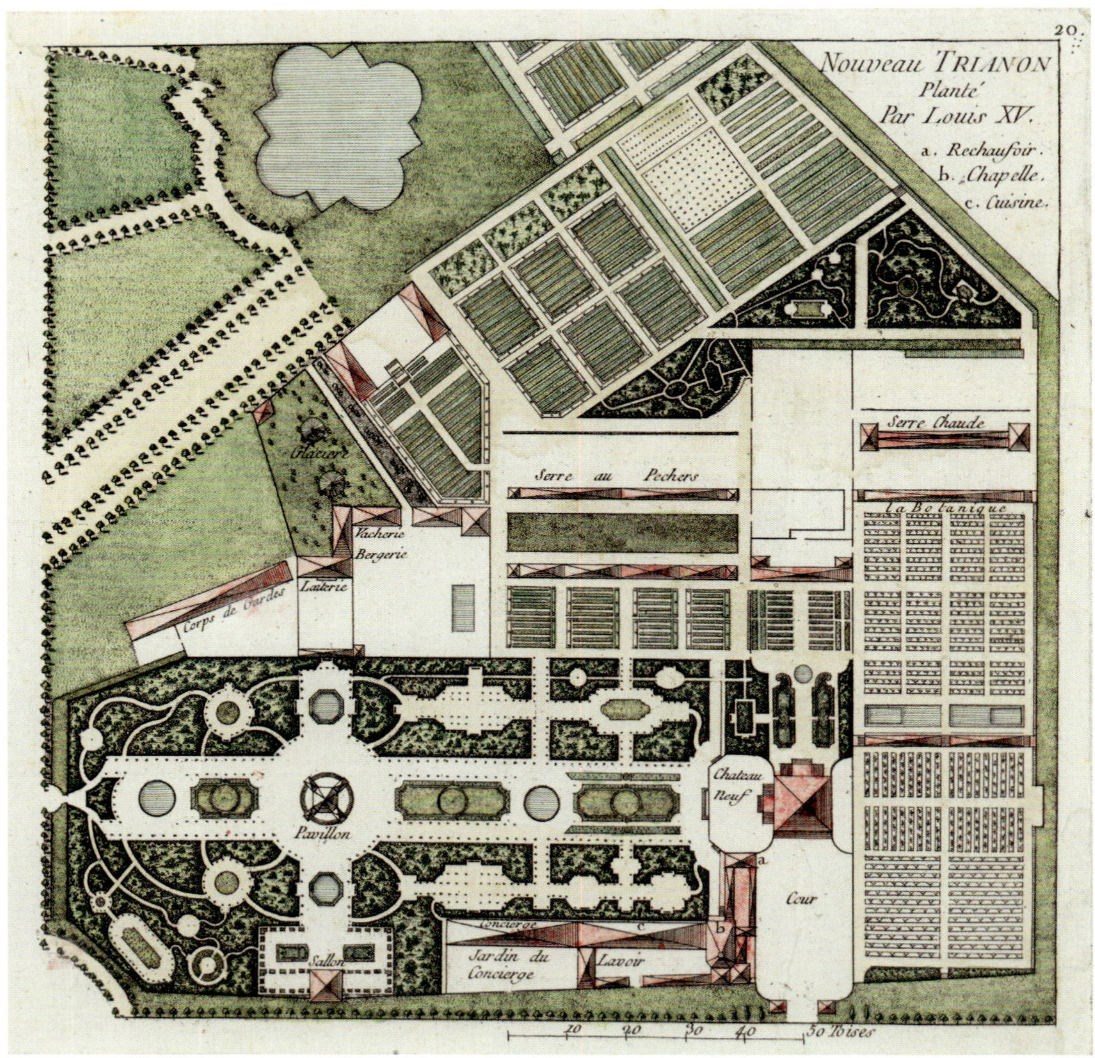

at Trianon, with different types of glasshouse along with an orchard equipped with hothouses for producing tropical fruits and vegetables, a flower garden for introducing new varieties, and a botanical garden exhibiting a collection of 4,000 plants with, among other things, a pool for aquatic plants.

The botanical garden at Trianon was first laid out in 1756, extended in 1761 and then completed in 1766 with the erection of a new large Dutch hothouse. In 1759 Bernard de Jussieu, a physician and botanist who had worked as a demonstrator at the King's Garden in Paris, aided by Michel Adanson, a traveller and botanist who had spent six years in Senegal from 1748 to 1754, arranged the plants according to his own scheme of classification, eschewing the earlier classification of the Swedish botanist Carl Linnaeus.

Philibert Commerson, a naturalist and member of Louis-Antoine de Bougainville's voyage of circumnavigation in 1766–69, brought the coco de mer to the estate of Trianon from the Île de France (Mauritius). Likewise, in 1761 Jean-Baptiste Fusée-Aublet, the apothecary-botanist of the French East India Company, sent Louis XV

ILL. 37
Plan of the new Trianon gardens planted for Louis XV, a plate from *Jardins anglo-chinois* (Anglo-Chinese Gardens), published by Georges-Louis Le Rouge (Paris, 1776–88), vol. 6, *c.*1778
Musée national des châteaux de Versailles et de Trianon. Object no. INV.GRAV 655

ILL. 38
Botanical study of *Drymonia coccinea* ('Besleria coccinea'), by Le Maire, as reproduced in Buc'hoz's *Jardin d'Eden*
Private Collection

plants and a 'rose butter' – a balsam whose scent the king came to love – made from damask roses and a distillation of his own invention. The naturalist abbé René Gallois sent a tea tree and a camellia from China.[3]

The presence of a large number of plants from Guiana at Trianon is one of the lasting effects of the Kourou expedition, which later became known as the 'disastrous Kourou expedition'. As the Seven Years War (1756–63) had resulted in France losing Canada, Louisiana and some islands in the Antilles under the Treaty of Paris signed on 10 February 1763, the duc de Choiseul, then Secretary of State for War and the Navy, launched the project of establishing a French colony in

Vanilla flore viridi & albo,
fructu nigrescente Plum.
cum flore.

Jasminum Arabicum
Lauri folio, cujus semen
apud nos Café dicitur,
Ac. R. Paris.

Ricinoides arbor,
Americana,
folio multifido.
I. r. h. 656.

ILL. 39
Botanical studies representing species grown at or sent to the Trianon gardens, the King's Garden, or encountered on French voyages, by Claude Aubriet, *c.*1700. Clockwise from top left: *Vanilla* – vanilla ('Vanilla flore viridi & albo ...'); *Coffea arabica* – coffee ('Jasminum Arabicum'); *Jatropha multifida* – coral plant ('Ricinoides arbor, Americana'); *Punica granatum* – pomegranate
RHS Lindley Collections, London. Object no. A/CAu/V/8; A/CAu/V/1; A/CAu/V/5; A/CAu/V/2

Guiana.[4] A note of royal assent from the king dated 1 April 1763 named, among others, a certain Le Maire, *dessinateur* (draughtsman) of the new colony. In Guiana, Le Maire drew plants brought to him by his host, Paul Demontis, a colonist who had settled in Roura, south of Cayenne. His drawings, some of which are signed, formed the *Catalogue des plantes de Monsieur d'Orcy*, now held in the French National Library. Forty-five of them were reproduced by Buc'hoz in the *Garden of Eden* (Ill. 38).[5]

On his return from Mauritius in 1762, Fusée-Aublet was also stationed in Guiana from 1762 to 1764 to raise the status of this new colony. He sent plants and seeds 'to enrich His Majesty's glasshouses'.[6] The governor of Guiana, Étienne-François Turgot, who returned to France in 1765, brought vanilla and a cacao tree for the hothouses at Trianon (Ill. 39). In 1781 the botanist Louis-Claude Richard, grandson of the king's gardener Claude Richard, was sent to Guiana by Louis XVI 'to enrich the Queen's botanical collection at the Petit Trianon'.[7]

Although Queen Marie-Antoinette abandoned the exploitation and scientific presentation of the plant kingdom that Louis XV had established at Trianon and ordered the demolition of the hothouses, the orangery and the botanical garden, the collections of plants patiently maintained by Claude Richard and his son Antoine survived in a different form. Buc'hoz's *Garden of Eden* is precious evidence of these collections originating mainly from North America, Guiana and the Indian Ocean.[8]

NOTES

1 Lamy 2010.
2 Lamy 2010–11.
3 Lamy 2023b.
4 Godfroy 2011.
5 Lamy 2023a.
6 Fusée-Aublet quoted in Lamy 2023a, p. 75.
7 Marquis de Castries to the baron de Bessner, 5 September 1781, Archives nationales d'outre-mer, Aix-en-Provence, COL C14 53, fol. 130.
8 Because the gardens of Versailles and Trianon no longer have plant collections in the strictest sense, there is no curator with specific responsibility for the gardens. The only curator who deals with the gardens is the curator for the sculpture collections.

9. The Clock of the Creation of the World

STÉPHANE CASTELLUCCIO

The commission for the *Pendule de la Création du monde* (Clock of the Creation of the World) came about in the context of the rivalry between Britain and France in India and the Indian nawabs' opposition to the authority of the Mughals (Ill. 40). Joseph François Dupleix, governor-general of the French colonies in India from 1742 to 1754, sought a strategic alliance with the nawab of Golconda, Salabat Jung (then known in France as Salabetzingue) (Ill. 41). Golconda was a central Indian state whose dependencies included Masulipatam (now Machilipatnam), an important centre for the production of calico – printed cotton fabrics that were much admired in Europe (Ill. 42). This alliance would have enabled the French East India Company to extend its territorial and economic influence in the centre of the Indian subcontinent. In return, Dupleix would have supported the nawab in his opposition to the Mughal.[1]

With the prospect of this alliance in mind, in the early 1750s Dupleix ordered from France an exceptional clock to be presented as a diplomatic gift to the nawab of Golconda. At that time, clocks were considered to be the most sophisticated and therefore the most prestigious devices ever designed. This astounding object was intended to impress the recipient by its size, the originality of its subject and shape, and its complex technology as evidence of the skill of French artisans and scientists. Dupleix approached the engineer Claude-Siméon Passemant, who was famous for having designed an astronomical clock that displayed the time, date, day, month and year, the phases of the moon and the positions of the planets (Ill. 43). Passemant had presented the mechanism to the Royal Academy of Sciences in 1749 and then to Louis XV at the royal palace of Choisy in 1753. The king, who had a great interest in the sciences and the arts, acquired it and ordered a gilded bronze case from the Caffieri brothers. When the case was completed, the timepiece took its place in the so-called cabinet de la pendule (clock cabinet) of the Petit Appartement du Roi at Versailles in 1754.

The title and ambitious subject matter of the clock commissioned by Dupleix, the creation of the world, announced the celestial phenomena to be displayed by its decoration and various mechanisms. At the top, a shining sun banishes the darkness, which is represented in the form of clouds. Its centre is composed of a dial showing the day and the month as well as the hour. In the middle, among the silver clouds, a globe spinning on its own axis presents the various phases of the moon. To the right, a dial reveals the orbits and positions of the six planets known at the time, revolving around the sun according to the Copernican system.

ILL. 40
Pendule de la Création du monde (Clock of the Creation of the World), designed by Claude-Siméon Passemant, 1754
Musée du Louvre, Paris, département des Objets d'art, VMB 1036-1, dépôt du musée national des châteaux de Versailles et de Trianon, 2011

ILL. 41
Posthumous marble bust of Joseph François Dupleix, by Charles-Antoine Bridan, 1787
Musée national des châteaux de Versailles et de Trianon. Object no. MV 1905

In the lower part, held in a patinated bronze rock surrounded by raging waves, sits a terrestrial globe engraved with the different countries. This globe rotates on its axis in 24 hours and the inclination of the poles varies according to the seasons. A long ray of sun indicates midday in the countries situated on the same latitude.[2]

To contain this amazing mechanism, the truly virtuosic creator of the case transformed the bronze into ethereal clouds, foaming waves and massive rocks. This interplay of forms and colours achieved by the mastery of the casting and engraving of the bronze, the gilding, the silvering and the dark patina of the metal creates an ensemble that provides a magnificent spectacle for the eye while the dials and globes make a deep impression on the mind.

A masterpiece of technology and beauty, this clock was presented to Louis XV at the Grand Trianon in the grounds of Versailles by Passemant himself on 6 February 1754. The sovereign was impressed by the scientific and technical achievements involved and asked for it to remain at Trianon until his next visit. Because of the snow, the clock had been transported by sledge, but it was so sturdily built that it was able to withstand the journey and the low temperature and functioned without problems.[3]

ILL. 42
Map of South Asia, engraved by François Desbruslins, eighteenth century
Bibliothèque nationale de France, département des Cartes et plans, Paris. Object no. GE AF PF-38 (77)

KEY
A. The Kingdom of Golconda
1. Masulipatam (Machilipatnam)
2. Pondicherry (Puducherry)
3. Chandernagore (Chandannagar)

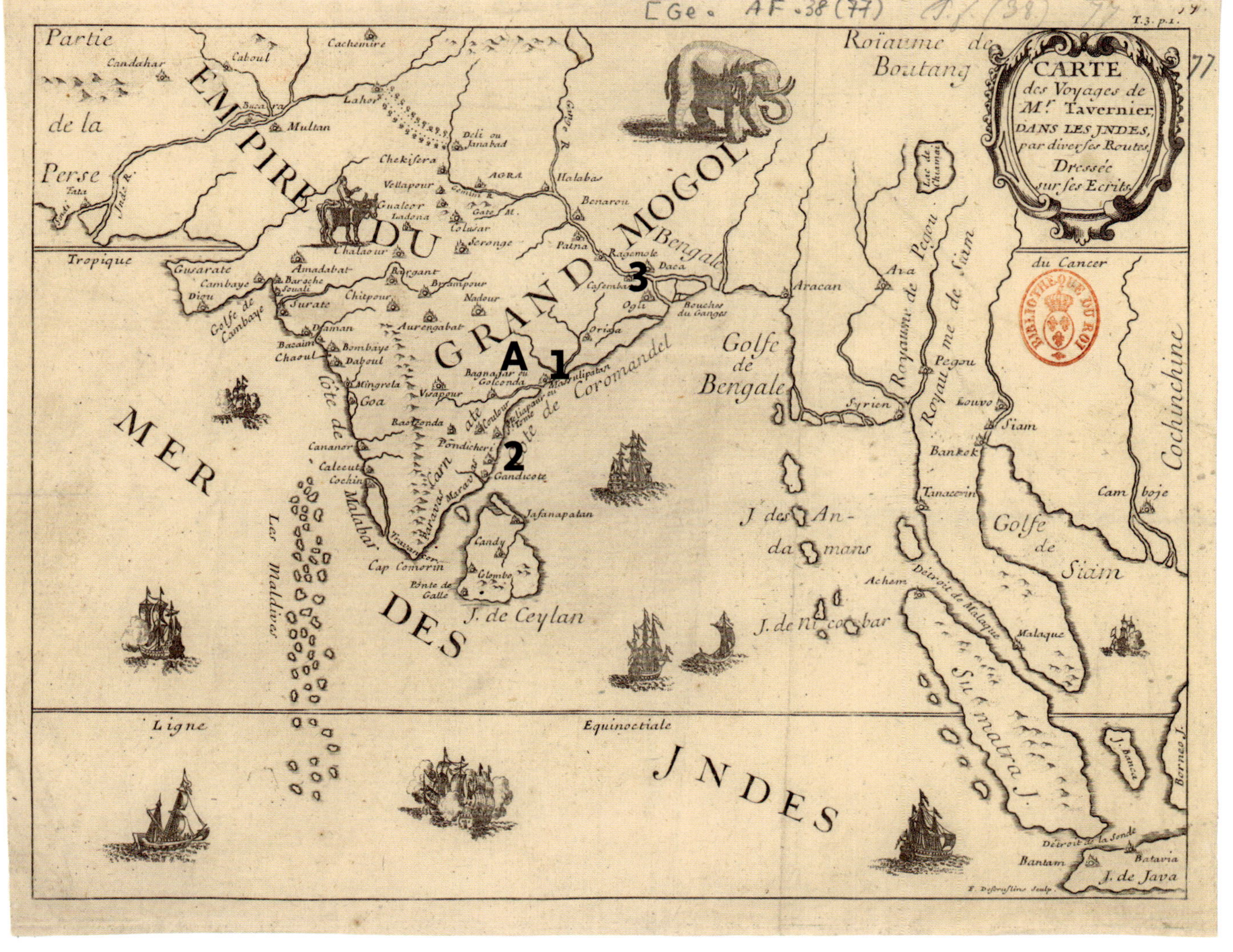

ILL. 43
Astronomical clock owned by Louis XV, designed by Claude-Siméon Passemant with clockwork by Louis Dauthiau and bronzework by Jacques and Philippe Caffieri, 1749–53
Musée national des châteaux de Versailles et de Trianon. Object no. VMB 1037

The clock then returned to Paris, where its mechanism was stopped before it was packed up and sent to India aboard the vessel *Diane*. In the meantime, Dupleix had been recalled to France, where he landed on 28 June 1755, in order to report to the management and shareholders of the French East India Company who disagreed with his Indian policy. In the end they appointed a new director in India and did not pursue the policy of alliances with local rulers introduced by Dupleix.[4] Agreements of this kind might possibly have avoided the British takeover of the French trading posts in India during the Seven Years War (1756–63). Under the peace treaty signed in 1763 only Pondicherry (Puducherry) was ceded back to France.

Following the directors' change in strategy, the clock was taken back on board in Pondicherry for return to France in October 1755, without having been offered to the nawab of Golconda. As it had not been ordered by the East India Company, it remained the property of the Dupleix family. The clockmaker Antide Janvier bought it when it went on public sale at the auction house of the Hôtel de Bullion, on 4 October 1791, and subsequently sold it to the French government in 1796.[5]

The Clock of the Creation of the World was as complex as the one acquired by Louis XV. It was considered a very rare object because of its history, its design and its intricacy, but nobody knew what to do with it after the French East India Company's change of policy. The case and mechanism bear witness to the complicated history of the company and its strategic errors, the scientific and technical expertise of Passemant, and the virtuosity of the creator of the case, who succeeded in transforming bronze and silver into the ancient elements of air, earth, fire and water in order to make a work of exceptional quality.

NOTES

1 Raynal 1774, vol. 2, pp. 114–19; Haudrère 2005, vol. 2, pp. 713–37.
2 Anon. 1754b, pp. 152–56; Anon. 1754a, pp. 105–6.
3 Castelluccio 2024, p. 123.
4 Raynal 1774, vol. 2, pp. 125–28; Haudrère 2005, vol. 2, pp. 738–45.
5 Bimbenet-Privat, Durand and Dassas 2014, pp. 280–81, n. 94; Carlier and Delalex 2022–23, p. 250, n. 114.

10. A Portrait of Emilie du Châtelet (1706–1749)

PATRICIA FARA

ILL. 44
Portrait of Gabrielle-Emilie Le Tonnelier de Breteuil, marquise du Châtelet, after 1733
Private collection Château de Breteuil – France

To succeed in science during the eighteenth century, a woman had to be ambitious, determined and very clever. Emilie du Châtelet certainly ticked all those boxes, but she also benefited from the advantages of possessing a large fortune, an accommodating family and a devoted lover: Voltaire. He quipped that she 'was a great man whose only fault was being a woman', and her portrait indicates how she pursued conventionally masculine interests while devoting much attention to maintaining outward appearances (Ill. 44).[1]

To pose for this oil painting, du Châtelet wore an elaborate costume and careful make-up, her hair fashionably pulled back from her forehead and probably powdered. Her sumptuous dress is edged with blue and yellow, her two favourite colours, which she chose for decorating her bedroom and its matching basket housing her puppy. While one hand supports her head in a classical pose of thought, the other clasps some dividers, a traditional symbol of male mathematicians but also probably a reference to the family legend that, at the age of three, she rejected a pair disguised as a doll by tearing off its clothes and working out for herself how to draw a circle. Behind her left shoulder are the rings of an armillary sphere, a modernised version of an ancient astronomical instrument for representing the heavens.

Du Châtelet gazes dreamily out at the viewer, as if momentarily interrupted in her contemplation of the open tome in front of her, an unrealistically large version of Isaac Newton's *Principia Mathematica* (1687). Although his great work on gravity had proved insurmountably challenging to all but the most eminent scholars, her translation from the original Latin remains the only complete one into French (Ill. 45). Unfortunately, the portrait is undated, but was almost certainly painted after 1733, when she began devoting herself to Newtonian mathematics. It is often attributed to Maurice Quentin de La Tour, but although he did depict several aristocratic women engaged in intellectual activities, he usually worked in pastels rather than oil.

The precocious daughter of indulgent parents, du Châtelet read voraciously and reportedly spoke six languages by the time she was 12 years old. As an adult in Paris, she led an unconventional life that included serious intellectual activities interspersed with dancing and gambling. Her husband, the father of her first three children, was often away pursuing military affairs, leaving her free to study Newtonian mathematics with private male tutors. By adopting English scientific ideas and discarding the approach of France's hero René Descartes,

du Châtelet made a decision that was both academically and politically controversial. Emotionally, her most important commitment was to Voltaire. She supported him financially, and when she was almost 30, agreed to join him permanently at her Cirey estate.

The couple entertained lavishly, but were also workaholics: they accumulated a massive library, carried out scientific experiments and wrote prolifically. The smaller volumes shown in her portrait may indicate the books and papers that they produced during almost 15 years living and working together. One of the first to appear, in 1738, was *Elémens de la philosophie de Neuton* (translated colloquially, *A Beginner's Guide to Newtonianism*), which was ostensibly by Voltaire but in fact a collaborative effort. He may have penned the words, but as his foreword stressed,

ILL. 45
Emilie du Châtelet's 1749 manuscript translation of *Philosophiae Naturalis Principia Mathematica* (Mathematical Principles of Natural Philosophy), by Isaac Newton (London, 1687), showing his First Law of Motion
Bibliothèque nationale de France, département des Manuscrits, Paris. Object no. Français 12266

ILL. 46
Frontispiece from *Elémens de la philosophie de Neuton* (Elements of Newtonian Philosophy), by Voltaire (Amsterdam, 1738)
Bibliothèque municipale de Versailles. Object no. F A in-8 A 139 a

the content owed much to her: 'The solid study that you have made of several new truths and the fruit of considerable work, are what I am offering to the Public for your glory'.[2] The frontispiece features du Châtelet towards the top right holding a mirror, not for admiring herself, but to identify her as the goddess of truth (Ill. 46). In a common Enlightenment motif, heavenly light of understanding streams through Newton to be reflected down onto Voltaire. Dressed as a classical scholar wearing a laurel wreath, he is assiduously transcribing her wisdom in his role of philosophical scribe.

Deliberately styled for a popular audience, their *Elémens* is often said to have been key to converting the French nation to Newtonianism. But du Châtelet probed deeper: not content with describing nature and its operations, she also explored the metaphysical foundations of science. Two years later, her *Institutions de physique* (Foundations of Physics) sought to reconcile divergent aspects of

ILL. 47
Frontispiece from *Institutions de physique* (Foundations of Physics), by Emilie du Châtelet (Amsterdam, 1741 edition)
The Linda Hall Library of Science, Engineering & Technology, Kansas City. Object no. QC19. D85 1741

Europe's three main authorities – Descartes, Gottfried Leibniz and Newton – represented by crudely drawn heads at the top of its frontispiece (Ill. 47). Beneath them, the author gropes through the dark clouds of ignorance, aspiring to reach the naked, radiant goddess of truth. On the ground below, allegorical female muses clutch instruments – including another armillary sphere – to symbolise various scientific disciplines, even though in reality women were excluded from participating in the generation of knowledge. Working in secret to avoid mockery, du Châtelet published anonymously, although her identity was soon revealed and her talent praised.

As implied by the portrait's enlarged representation of the *Principia*, translating it was a colossal undertaking, which she made still more onerous by including extensive commentaries. In addition to her literal translation of Newton's prose, du Châtelet provided more accessible explanations of his arguments with clear examples. She transformed his geometrical expositions into modern calculus, and also summarised the latest mathematical research and experimental vindications of his theories. On top of all that, she struggled through a difficult pregnancy, working 18 hours a day to complete her ambitious, self-imposed task. Although she did finish it, she died a couple of weeks later, never knowing that her book appeared posthumously, astutely timed to coincide with the reappearance of Halley's comet in 1759.

One acquaintance remarked that du Châtelet put 'so much effort into appearing what she is not, that one no longer knows who she really is'.[3] Some three hundred years later, her portrait reveals only what this exceptional woman wanted posterity to perceive.

NOTES

1 My major biographical source is Zinsser 2006. See also Fara 2004, pp. 88–105. Letter to Frederick II of 15 October 1749: Besterman 1953–65, vol. 17 (1956), p. 188.
2 Voltaire 1738, pp. 9–10.
3 Madame du Deffand quoted in Poirier 2002, p. 237.

11. Abbé Jean-Antoine Nollet's Spectacle of Science

JEAN-FRANÇOIS GAUVIN

'Abbé Nollet is ruining me', wrote Voltaire in 1738, albeit quickly adding it was easier to find money than a *philosophe* like the abbé to build a complete cabinet of physics.[1] Abbé Jean-Antoine Nollet began his celebrated career as an *artiste*, or a craftsman endowed with *esprit*, a learned and active mind capable of creativity and invention. He joined the Société des Arts (Society of Arts) in Paris in 1728, a company of like-minded *artistes* dedicated to improving the economic status and the administration of the state. There, Nollet established his first network of influential contacts, such as the naturalist and academician René-Antoine Ferchault de Réaumur to whom he became an assistant. Within a decade, Nollet confirmed he was exceptionally gifted with his hands as well as a genuine natural philosopher; he was promptly elected in 1739 to the Royal Academy of Sciences, the pinnacle of French science in the eighteenth century.

From the 1730s onwards, Nollet launched what would become a lucrative and socially prestigious 'industry' focused on the manufacturing and selling of scientific instruments and their use in public and private lecture demonstrations.[2] In Paris, the experimental courses took place in three main areas, depending on the target audience: the Left Bank, where one would find the educated and bourgeois Parisians; the Boulevard du Temple, which served as a centre for popular entertainment; and the Palais-Royal area, where the upper class and economic elite attended courses and bought instruments. Taking the city by storm, this novel type of sensational entertainment was somewhat frowned upon by Voltaire, who feared that poetry was no longer à la mode in Paris. Instead, people were 'meddling with reason' and turning into geometers or physicists.[3] Such lessons became so popular that Emilie du Châtelet, mathematician extraordinaire, translator of Newton and Voltaire's muse, apparently reported that only carriages of duchesses, peers and pretty women showed up at Nollet's door. In England, too, the artist Joseph Wright of Derby suggested the growing interest in the eighteenth century for the spectacle of science when he painted a natural philosopher demonstrating to an elite crowd the effects of a vacuum on a cockatoo (Ill. 49). Whether one was an aristocrat conversing in a salon, a commoner standing at a street corner or a bourgeois visiting an instrument-maker's workshop in Enlightenment Paris, all witnessed astonishing experiments and skilled performances exhibiting revolting smells, multicoloured optical displays and deafening booms.

Delightful and amusing physics, with its myriad of tricks, performances and especially its power of replication, provided a sense that natural phenomena could

ILL. 48
Air pump, designed by and made for abbé Jean-Antoine Nollet, *c.*1758
Musée des Arts et Métiers – CNAM, Paris. Object no. 07516-0000-

be revealed to one and all. It contributed to the transformation of the public sphere pertaining to the realm of politics, education and consumerism. Nollet's rising celebrity in the 1740s qualified him as a regular and esteemed guest at the royal court in Versailles. In 1746, for example, he demonstrated on 180 unsuspecting royal guards standing in Versailles's Hall of Mirrors how powerful electrical shocks produced by the newly discovered Leyden jars were. A decade later, in 1758, the king employed him as tutor of physics and natural history to the Enfants de France (the sons and grandsons of the sovereign). Nollet supplied a collection of scientific instruments for their education, including an air pump, one of the most iconic philosophical instruments since the seventeenth century, which allowed for a variety of experiments as depicted by Wright of Derby, as well as a cycloidal track, which demonstrated that the shortest time between two points is given by the track made from the shape of a cycloid (not a straight line or circular arc) (Ill. 48, 50). Nollet's reputation extended outside of Paris as well, and across the Alps. As the foremost 'electrician' in Europe until the early 1750s (a title he eventually lost to Benjamin Franklin), Nollet visited Italy for the second time in

ILL. 49
An Experiment on a Bird in the Air Pump, by Joseph Wright of Derby, 1768
The National Gallery, London. Object no. NG725

ILL. 50
Cycloidal track, designed by and made for abbé Jean-Antoine Nollet, 1759
Listed as a historical monument on 17 March 1971, PM78000690
Lycée Hoche, Versailles
On long-term loan to Musée Lambinet, Versailles. Object no. D.90.2.4

1749 – a decade earlier he had been received by Charles Emmanuel III, Duke of Savoy and King of Sardinia, to whom he taught experimental physics. The 1749 trip was advertised as an impartial assessment of the medical 'miracles' documented by some Italian savants using electricity together with therapeutic 'medicated tubes'. Nollet not only debunked these so-called 'truths', but also met with the Pope and conducted industrial espionage on behalf of France regarding the lucrative and coveted silk industry. Upon his return, he continued to expand his French experimental physics 'empire' by accepting several prestigious appointments: in 1757 at the artillery school of La Fère, where he offered instruction in algebra, geometry, hydraulics, physics and chemistry; at the engineering school of Mézières in 1761; and at the artillery school of Bapaume in 1765.

Savant, teacher, courtier, *artiste* and spy, Nollet was an ideal member of the Republic of Letters, an intellectual space and network composed of (mostly) men and women exchanging news and courtesies and arguing about natural facts and the nature of knowledge. The first holder of a chair in experimental physics at the

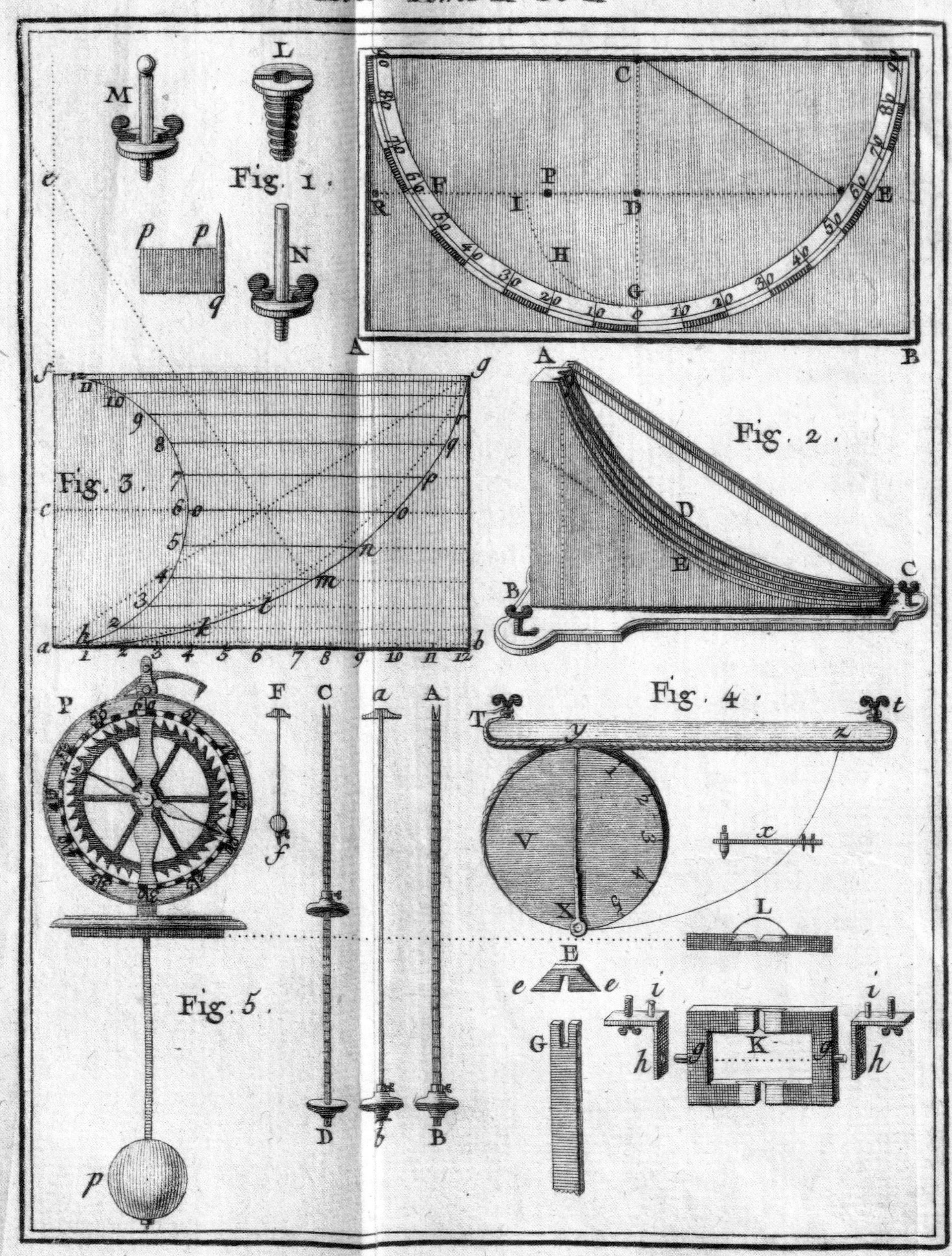
Fig. 1.
Fig. 2.
Fig. 3.
Fig. 4.
Fig. 5.

ILL. 51
Plate including a diagram of Nollet's cycloidal track (Fig. 2), from his book *L'Art des expériences* (The Art of Experiments) (Paris 1770), vol. 2
Bibliothèque municipale de Versailles. Object no. Res in-12 A 15 e

Collège de Navarre (1753), Nollet published a series of six volumes, *Leçons de physique expérimentale* (Lectures in Experimental Philosophy) (1748–64), that was often re-edited and translated during his lifetime. With the instruments Nollet was keen to provide, it became possible to replicate for oneself, or for a whole group seeking fashionable entertainment, the same lessons Nollet delivered throughout France. Such was his success and fame that a few years after the suppression of the Jesuit order in France in 1763, Nollet explained to one correspondent he was receiving so many requests from teachers concerned with teaching experimental physics in newly founded colleges that only the publication of his complementary book on the manufacture of instruments would solve this epistolary chore. His book became the three-volume set *L'Art des expériences* (The Art of Experiments) published in 1770, the year Nollet died. By carefully explaining how to manufacture his array of scientific instruments, Nollet hoped to create an enduring pedagogical movement, and, of course, to secure a durable legacy (Ill. 51). It worked for a few decades after he passed away. Around 1776, Joseph-Aignan Sigaud de Lafond, Nollet's true heir, furnished the Collège des Godrans in Dijon with a set of instruments (for which he was paid 100 bottles of Burgundy). In Paris about 1790, the engineer and optician Hæring, working at the sign of the *l'Aigle d'or* (Golden Eagle), No. 63 at the Palais du Tribunat, near the Café de Foi, advertised electrical instruments made according to Nollet's specifications.

Although Nollet's name today might not be well known to the public, his heritage is: it has been recaptured and reinvented by science museums and institutions. Next time you see someone's hair rise straight up in contact with an electrostatic machine, think of Nollet and his carefully orchestrated Enlightenment spectacle of science.

NOTES

1 Voltaire to Nicolas Claude Thieriot, 27 [October 1738], D1640, in Besterman 1968–77, vols 85–135.
2 I use in this text the generic term of 'scientific instrument' to describe Nollet's manufactured devices. For more on the evolution of this term, see Taub 2019.
3 Voltaire to Pierre-Robert Le Cornier de Cideville, [16 April 1735], D863, in Besterman 1968–77, vols 85–135.

12. Madame du Coudray's 'Machine'

SARAH BOND

Inhabiting the uncanny territory between dressmaker's dummy and anatomical Venus, this extraordinary set of objects is the only known surviving example of an obstetric teaching model that transformed midwifery instruction in eighteenth-century Europe (Ill. 52).

ILL. 52
La Machine: obstetric teaching model, designed by Angélique Marguerite Le Boursier du Coudray, c.1778
Musée Flaubert et d'histoire de la médecine, Réunion des musées, Métropole Rouen Normandie, Rouen. Object no. 2004.0.58.1, 2004.0.58.3 and 2004.0.58.7

The life-sized lower torso sits atop a wooden frame mimicking a birthing chair. Canvas flaps covering the abdomen peel back to reveal the uterus and its surrounding organs faithfully recreated in linen, hide and silk taffeta dyed various shades of pink. Each part is numbered, and a series of straps and drawstrings allows for adjustments to simulate advancing labour.

A doll the length of a full-term baby rests between the mannequin's open thighs, still attached to its mother via a cloth umbilical cord. X-rays reveal the pair's internal composition: a real human pelvis provides structure for the torso, while the bony parts of the child are made from rigid materials padded with cotton to replicate soft tissues (Ill. 53).

The attention to detail is remarkable. Were you to stroke the top of the infant's head, you would feel the gentle dip of the fontanelle, and each of its vertebrae – crafted from starched, concertinaed fabric – can be individually identified by touch. The mouth is open for good reason: it enabled insertion of a finger to demonstrate how to safely deliver the head during a breech (bottom first) birth.

This version was a 'dry' model; others featured sponges that could be soaked in clear or dyed liquids to stage the release of amniotic fluid and blood. Accessories include a seven-month fetus and a set of twins at approximately five months' gestation with shared placenta, one side of which is meticulously embroidered in red and blue to represent the arteries and veins.

Resembling its present-day equivalents in all but materiality, this ingenious device was the creation of Angélique Marguerite Le Boursier du Coudray, the midwife commissioned by Louis XV to train a generation of women and male surgeons in the art of delivery (Ill. 54).

By the mid-eighteenth century, alarming reports from the countryside of many thousands of babies dying during or shortly after birth heightened concerns that the French population was being decimated at home as well as on the battlefields of recent conflicts, including the Seven Years War (1756–63). Boosting the number of royal subjects thus became a priority for the monarchy, whose prestige depended on securing the state's future military capacity.

In October 1751, du Coudray, then head midwife at Paris's oldest hospital, the Hôtel-Dieu, returned to her native province of Auvergne where the devastating consequences of unskilled midwifery were immediately apparent. Births were typically attended by matrons: village women with plenty of experience but little to no medical training, who were paid in kind for their services. Associating prolonged deliveries with complications, they often attempted to hasten the process. If the child could not be dislodged by tugging an emerging part or pressing on the mother's abdomen, they might encourage her to walk or jump – advice that, if mistimed, could result in the baby's death by strangulation.[1] In the worst-case scenario, when a stuck infant was deemed beyond saving, household implements were used to remove it piecemeal

Appalled by the stories of rural mothers, many of whom were left permanently injured or infertile by botched deliveries, du Coudray devoted the next decade to teaching in the region. Lessons were provided free of charge to peasant girls and women with no previous experience. Given the scale of the challenge, the midwife recognised that to have any impact she would need to teach sizeable groups in a relatively short space of time – a far cry from the three-year, one-to-

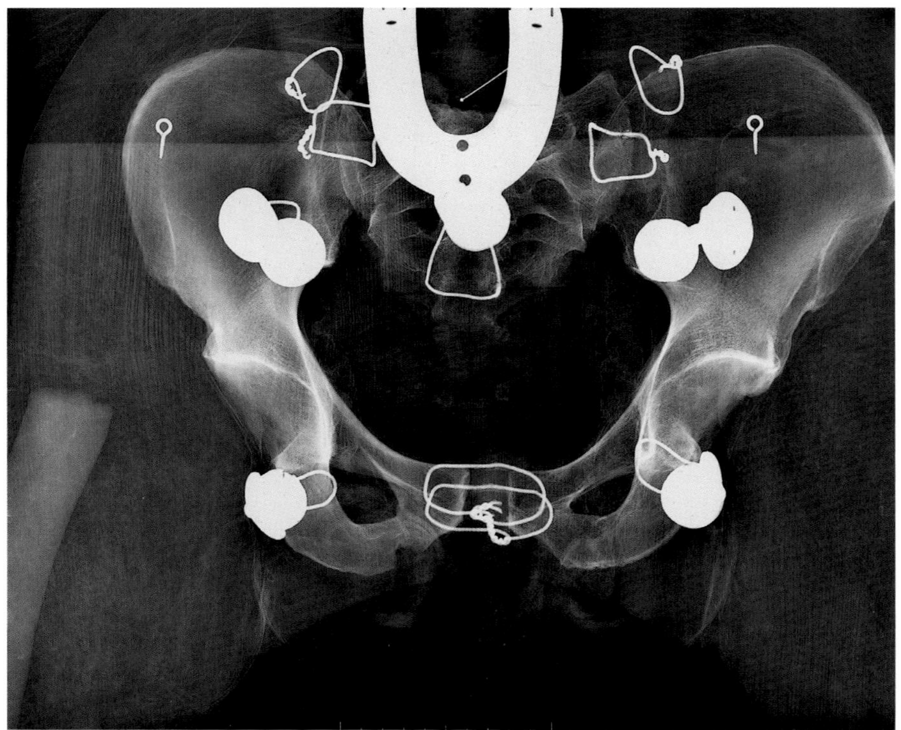

ILL. 53
Radiograph of the delivery mannequin
Musée Flaubert et d'histoire de la médecine, Réunion des musées, Métropole Rouen Normandie, Rouen

one apprenticeships she and her colleagues had undertaken in the capital. 'The only obstacle I found to my project', du Coudray declared, 'was the difficulty of making myself understood by minds unaccustomed to grasping things except through the senses'. To this end, she opted to make her 'lessons palpable' for her largely illiterate cohort 'by having them manoeuvre in front of me on a machine I constructed for this purpose'.[2]

While not the first anatomical model for demonstration, du Coudray's 'machine' was unparalleled in its sophistication and, crucially, allowed students to gain practical, hands-on experience in the mechanics of childbirth. As well as practising straightforward deliveries, they had the opportunity to enact rarer scenarios – abnormal presentations, twins, a retained placenta – that often had fatal consequences without deft intervention.

So successful was du Coudray's invention that in May 1756 it was given the seal of approval by the all-male Académie royale de chirurgie (Royal Academy of Surgery) in Paris. High praise from one of the reviewers, esteemed obstetrician André Levret, further enhanced her reputation in Versailles. The tipping point, however, came in 1759 with the publication of her *Abrégé de l'art des accouchements* (Summary of the Art of Childbirth). Devised to accompany du Coudray's lessons and reinforce techniques learnt on the machine, the textbook explicitly equated her patriotic mission to 'watch over the conservation of His Majesty's subjects' with victory in war.[3] It was an unpretentious manual written in French (rather than Latin), which used familiar terms and frames of reference – comparing cervical dilation to the size of a fish's open mouth, for example – to keep it 'accessible to those least schooled in this matter'.[4] The second edition featured

ILL. 54
Frontispiece with portrait of Angélique Marguerite Le Boursier du Coudray, from her book *Abrégé de l'art des accouchements* (Summary of the Art of Childbirth) (Paris, 1777 edition)
Science Museum Group, London. Object no. 1979-455/4

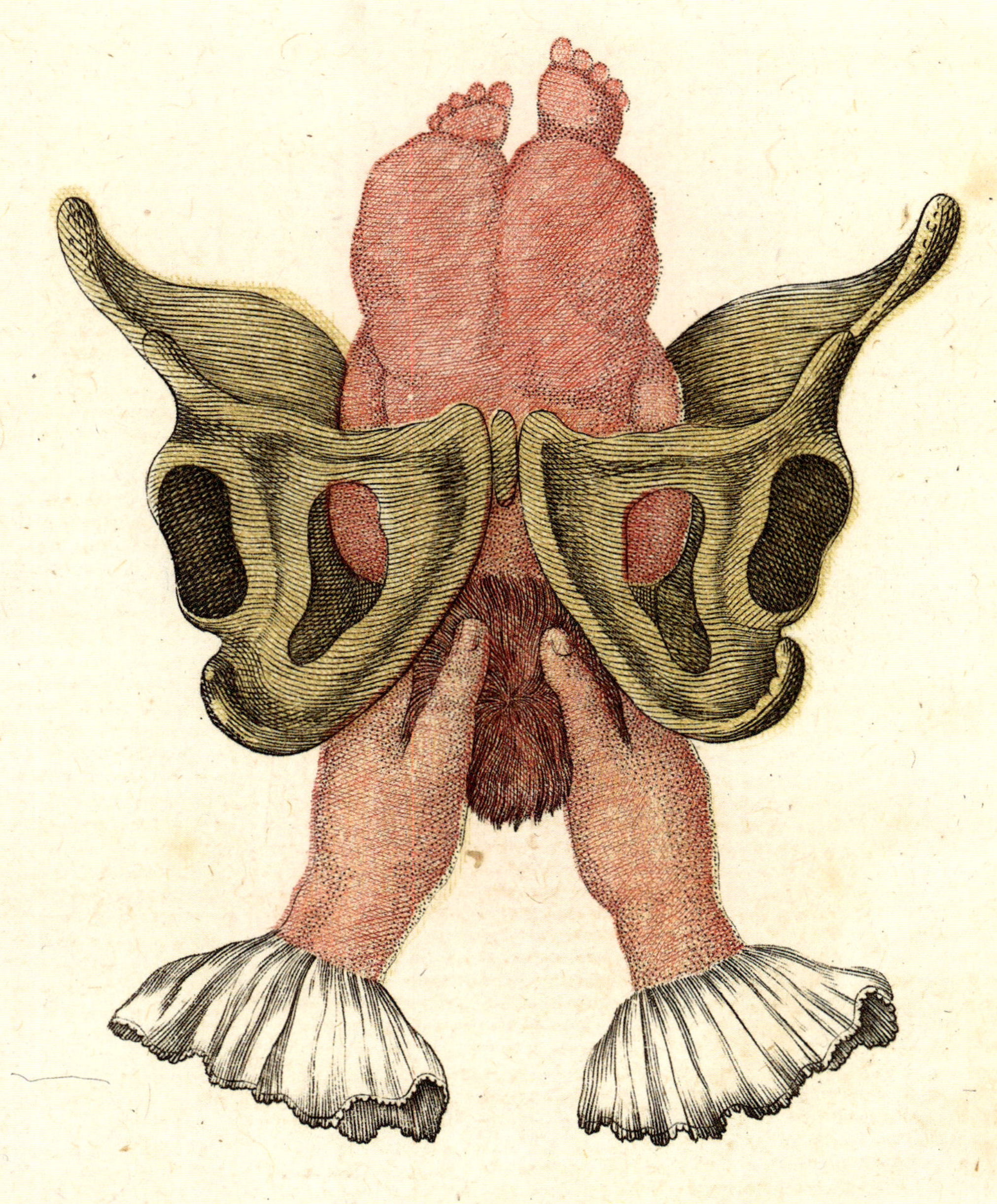

ette figure répresente l'Enfant qui vient naturèlement avec la posit
onvenable des mains aux deux côtés de la tête pour le tirer en bas

ILL. 55
A straightforward birth with a baby being delivered into the clean, waiting hands of a midwife, a plate from du Coudray's *Abrégé*
Science Museum Group, London. Object no. 1979-455/4

multicolour illustrations, never before used in an obstetrical text, to imprint vividly the fundamental manoeuvres in her students' minds (Ill. 55).

Before the year was out, du Coudray had secured a royal brevet: a decree granting her freedom to teach throughout France without encumbrance. She was 47 years old when she embarked on what would become her life's mission, traversing the country to deliver the first nationwide programme for midwifery education. The goal was to create 'a vast emergency squad' of skilled practitioners to reverse the perceived depopulation crisis.[5] Intensive courses took place over two months. Students, whom du Coudray referred to affectionately as 'mes femmes', were awarded a certificate – and later a copy of the *Abrégé* – upon completion to mark their achievement. Other incentives for would-be midwives included lifelong tax exemptions, free accommodation during their studies and immunity for their husbands from forced work on the construction of royal roads.[6]

Male surgeon-demonstrators were trained alongside the women on condition that they would continue to teach in each province after du Coudray's departure. Under this franchise-like arrangement she made her visits contingent on the purchase of multiple machines, including the most expensive silk version for deposition in the local Hôtel de Ville, where it would be permanently available for consultation. Du Coudray produced and sold hundreds of the models over the next quarter of a century, a strategy that resulted in an astonishing two-thirds of the nation's midwives being trained according to her method by 1786. Constituting 'nothing less than a revolution in medical pedagogy', the ripple effect created by this exceptional woman almost certainly contributed to the marked reduction in infant mortality and subsequent population boom of the latter eighteenth century.[7] Du Coudray herself was unequivocal: her pioneering machine would serve as 'a monument to humanity for the centuries to come'.[8]

NOTES

1 Le Lous, Baxter and Timoh 2022, p. 2.
2 Le Boursier du Coudray, *Abrégé*, 'Avant propos', pp. v–viii, translated and cited in Gelbart 1998, p. 60.
3 Le Boursier du Coudray, *Abrégé*, 'A Monseigneur Bernard de Ballainvilliers', translated and cited in ibid., p. 75.
4 Le Boursier du Coudray, AD C C319, fol. 24, cited in ibid., p. 16.
5 Gelbart 1996, p. 1003.
6 Gelbart 1998, p. 162.
7 Gelbart 1996, pp. 1000–1.
8 Le Boursier du Coudray, AD IV C1326, fol. 18, translated and cited in Gelbart 1998, p. 192.

13. 'A Fairly Good Likeness': Diderot and his Portraits

LUDMILLA JORDANOVA

The pursuit of knowledge was a serious business in *ancien régime* France. By circulating manuscripts and letters in addition to printed materials, it was possible to exchange ideas not just in formal institutions, such as academies, but among friends and associates, some of whom held deeply controversial views. Rulers and established elites took a keen interest in these matters too, sometimes acting as patrons to well-known thinkers and writers, especially those who enjoyed a measure of celebrity. Portraits, especially in the form of prints, often used as frontispieces to books, enabled the appearance of famous thinkers to be disseminated far and wide. As they had been for centuries, likenesses of prominent people were everywhere, including on coins and medallions, in the form of statues and busts, miniatures and canvases as well as prints using a range of techniques and available at prices to suit many pockets, from expensive, gorgeous mezzotints to small, ephemeral etchings.

The topics covered by authors were strikingly diverse in a setting innocent of the terms we now use to distinguish between types of knowledge and to police boundaries between academic fields. Natural history, natural philosophy and medicine along with crafts requiring technical know-how generated wide interest alongside literature, the arts and commentary about contemporary society. Denis Diderot is best appreciated as an intellectual dynamo, capable of engaging with a vast array of topics and turning his hand to many genres with wit, verve and a critical eye. As a controversial figure holding unorthodox views, his portraits are of considerable interest, not least because of his own responses to them. Curiosity about the natural and social worlds and about people went hand in hand and manifested themselves in the visual arts, including portraiture.

Although not the most portrayed of the *philosophes*, Diderot was depicted by a number of artists, some of whom he deemed friends.[1] A man of extraordinary vitality, he wrote novels, plays, treatises and dialogues as well as numerous articles in the great *Encyclopédie* of which he was the general editor. He had to tread carefully when it came to religion and morality; nonetheless, his ambitious venture reached many educated and curious people, providing them with information on an astonishing range of topics along with observations both sharp and wry on contemporary life. Diderot was also an art critic, although his commentaries on the so-called Salons, free exhibitions held in the Louvre by the Académie royale de peinture et de sculpture (Royal Academy of Painting and Sculpture), were not published in his lifetime, but circulated in manuscript form to a small

ILL. 56
Portrait of Denis Diderot, Writer, by Louis-Michel Van Loo, 1767
Musée du Louvre, Paris, département des Peintures. Object no. RF 1958

number of privileged people interested in the arts and in collecting. These commentaries, full of the digressions, speculation and flights of fancy found in much of his work, contain his response to the portrait by his friend Louis-Michel Van Loo and to other depictions of him that had been made by 1767, the year in which the portrait in question was publicly displayed (Ill. 56).[2]

In 1766 Diderot's then friend Jean-Baptiste Greuze had executed a chalk drawing that seems to have met with the sitter's approval, serving as the template for a number of engravings (Ill. 57). A head in profile immediately reminds viewers of coins and medals. As in the portrait by Van Loo, Diderot is wigless. The drawing can be seen as 'neoclassical' in its simplicity, not anchored to a specific time and place. The same can be said of the 1771 bust by Jean Antoine Houdon, where there is not a hint of clothing (Ill. 58). Diderot, a man of wide reading, huge curiosity and considerable learning, found such classicism appealing. The contrast with the Van Loo painting could hardly be starker. The subject is seated at a table, pen in hand, and glancing to one side as if setting his thoughts in order before

committing them to paper. The gesture of his left hand further suggests we are witnessing a fleeting phenomenon. His pose invites spectators to imagine they have interrupted him in mid flow, that he might turn at any moment and engage them in conversation. Van Loo provided Diderot with the standard accoutrements of a writer: paper, pen and ink. But the soft blue garment he wears over a loose white shirt is quite sumptuous, and unlikely to have been Diderot's own, as he protests: 'And then clothing so luxurious as to ruin the poor man of letters should the tax collector levy payment against his dressing gown'.[3] Writers were frequently portrayed in such attire, which to modern eyes hardly looks like a dressing gown, but in the period it commonly conveyed the sense of a man in an interior space designed for reflection and composition. Diderot was in his fifties at this point, and felt he was shown as too young and indeed too effeminate, even if it was 'a fairly good likeness'.

The exact date of production of the print based on Van Loo's painting is unclear; it generates a quite different visual effect to the delicate colours and softened features of the painting (Ill. 59). In the engraving, a bit bigger than an A4 sheet of paper, Diderot looks more his actual age. He spent six months in Russia in 1773–74, when the engraver, Benoît-Louis Henriquez, who had worked in Paris, was also there. Catherine the Great's intellectual and cultural enthusiasms are well known.

ILL. 57
Portrait of Denis Diderot, by Jean-Baptiste Greuze, 1766
The Morgan Library & Museum. 1958.3. Gift of Mr. John M. Crawford

ILL. 58
Terracotta bust of Denis Diderot, by Jean Antoine Houdon, 1771
Musée du Louvre, Paris, département des Sculptures du Moyen Age, de la Renaissance et des temps modernes. Object no. RF 348

She amassed a formidable art collection and manifested a genuine interest in ideas, including those explored by radical thinkers such as Diderot. It was precisely because his ideas were unconventional, challenging religious and political orthodoxies, that Diderot was careful about what he published, given his brief stay in prison in 1749. Although many of his works remained unpublished in his lifetime, his ideas and writings circulated informally, including through conversation. Like so many of his contemporaries, Diderot was a prolific letter writer and enjoyed the company of leading figures in mathematics, medicine and natural history through whom he was able to learn of and debate fresh discoveries and theories.

ILL. 59
Portrait of Denis Diderot, engraved by Benoît-Louis Henriquez after Louis-Michel Van Loo, 1767–1804
Musée national des châteaux de Versailles et de Trianon. Object no. INV.GRAV 2982

ILL. 60
1772 frontispiece from *L'Encyclopédie, ou Dictionnaire raisonné des sciences, des arts et des métiers* (The Encyclopedia, or A Systematic Dictionary of the Sciences, Arts and Crafts), edited by Denis Diderot and Jean le Rond d'Alembert (Paris 1751–72), text vol. 1
Wellcome Collection, London. Reference 46965i

The continuing significance of scribal culture, as is clear from the manner in which his Salon criticism was transmitted, reveals the vibrancy of intellectual life and the extensive exchanges that took place beyond the realm of print. That Diderot sometimes wrote in the form of a dialogue further reinforces the point, even if one such text, *D'Alembert's Dream*, described as the summa of his thoughts on the life sciences, was not published until 1830, having been composed in 1769.[4] 'Dialogue' is altogether too dry a word for the conversations it contains, which are so full of energy and imagination that they fly in many directions, exhilarating the reader and revealing just how embedded that phenomenon we call 'science' was in the cultural life of *ancien régime* France. This racy text touches on reproduction, medicine, sexuality, human physiology and the nature of life: all key areas for those who eschewed explanations that relied on the existence of a soul, a designing hand or a deity. Such writings sit side by side with the *Encyclopédie*, which celebrates crafts and technologies and was committed to truth and reason. A preparatory drawing for its frontispiece was exhibited at the Salon of 1765: Diderot largely approved of a depiction in which reason and imagination triumph over metaphysics and theology (Ill. 60). The conviction that widely disseminated knowledge of high quality would bring improvement to the human condition lay at the heart of the Enlightenment: Diderot shows that it was perfectly compatible with play, delight in sensual experience, and the embrace of speculation and ambiguity.

NOTES

1 Sahut and Volle 1984–85.
2 Diderot 1995, p. 19–21.
3 Ibid., p. 19.
4 Hoare 2022, vol. 1, p. 420; Diderot 1966.

14. A Chemistry Laboratory Model for Princely Education

ANNE-LAURE CARRÉ

This little model of a chemistry laboratory relates to both the tradition of collecting models and to the interest in science and technology in the late eighteenth century, which manifested itself in the public demonstrations of new machines and the fashion for scientific displays (Ill. 61).[1] At the time, Versailles possessed specially equipped rooms and collections of instruments for the scientific education of the young princes. A position of tutor of physics and natural history to the children of Louis XV was created and entrusted to the abbé Nollet; later, a second post would be created for the education of the future Louis XVI and his brothers.[2] Louis XVI was passionate about science and technology, particularly maritime engineering and cartography, and practised mechanics in his private apartments.

In 1782 Louis-Philippe, duc d'Orléans, one of King Louis XVI's cousins, commissioned 17 models of artisanal workshops to form a 'collection of models of the arts and crafts and manufacture' that was installed in his residence in the Palais-Royal in Paris, next to the art galleries, and the gemstones and natural history cabinets.[3] The models were ordered from the Périer brothers, who were protégés of the duc d'Orléans, but it was later revealed that they had been made by the young François-Etienne Calla before he achieved fame as an iron founder, mechanic and boilermaker. Recent research based on a rediscovered document has established that another 'artist-mechanic', Mathieu Laveron – not as well known but no less interesting – had worked on creating them.[4]

These models would also have served to educate the children of the duc d'Orléans, including the future French king Louis-Philippe. Stéphanie-Félicité du Crest de Saint-Aubin, comtesse de Genlis, who was appointed tutor of the young male heirs in 1782, specifically mentioned their creation in her memoirs. An exceptional woman, her appointment to this post, normally strictly reserved for men, caused a scandal. She favoured new ideas about education with a focus on the experience of the senses, object-based learning and physical exercise. Her young pupils made furniture, gardened and visited factories and workshops. In October 1782 they went to the village of Chaillot, to see the new fire pumps built by the Périers from plans drawn up by Matthew Boulton and James Watt to raise water from the Seine and supply a network of fountains.

During the Revolution, the models were seized and from 1802 they were deposited at the Conservatoire des Arts et Métiers (National Conservatory of Arts and Crafts). The 13 that still survive today mainly represent *les arts du feu* (skills

ILL. 61
Model of a chemistry laboratory, made by François-Etienne Calla and Mathieu Laveron, 1783
Musée des Arts et Métiers – CNAM, Paris. Object no. 00131-0000-

requiring the use of heat): nail-making, locksmithing, casting metal, making metal pipes, and making pottery, faïence and porcelain. Two models are dedicated to chemistry: the making of aqua fortis (nitric acid) and the laboratory.

The source of their inspiration was undoubtedly the *Encyclopédie*, published in 17 volumes of texts and 11 volumes of plates and descriptions, between 1751 and 1772. This was the first French encyclopedia, modelled on the *Cyclopaedia* of Ephraim Chambers published in 1728 in London. After an unfortunate attempt at translation, the publisher André Le Breton called upon a society of men of letters, including the mathematician Jean le Rond d'Alembert and the writer Denis Diderot. In the context of changing ideas and political claims, the *Encyclopédie* became the symbol of the Enlightenment and its most remarkable editorial enterprise. The articles were not restricted to a glossary of terms; for example, regarding the mechanical arts, they formed an unparalleled body of descriptions of the processes and tools used. The entries for 'chemistry' and 'chemistry laboratory' were written by the physician and chemist Gabriel-François Venel, a regular contributor (more than 700 articles!) and determined to defend the position of his subject as an entirely separate science.[5] Louis-Jacques Goussier engraved no fewer than 25 illustrative plates in the volume *Recueil de planches sur les sciences, les arts libéraux, et les arts méchaniques avec leur explication*, which appeared in 1763.[6]

The model of the chemistry laboratory captures every detail of the drawing in the plate overleaf, but without people (Ill. 62). At first glance, the enormous chimneypiece dominates the scene; it meets the ventilation requirements emphasised in the instructions as well as providing a display stand for all the tiny pieces of laboratory equipment. The chimney hood is covered with shelves lined with three rows of receptacles and vessels that can be placed on the heat: stoneware pots, bowls, crucibles, long-necked glass flasks and alembics and earthenware retorts. The symbols inscribed along the edge are borrowed from the lower part of the same plate and identify the elements: the fixed and volatile

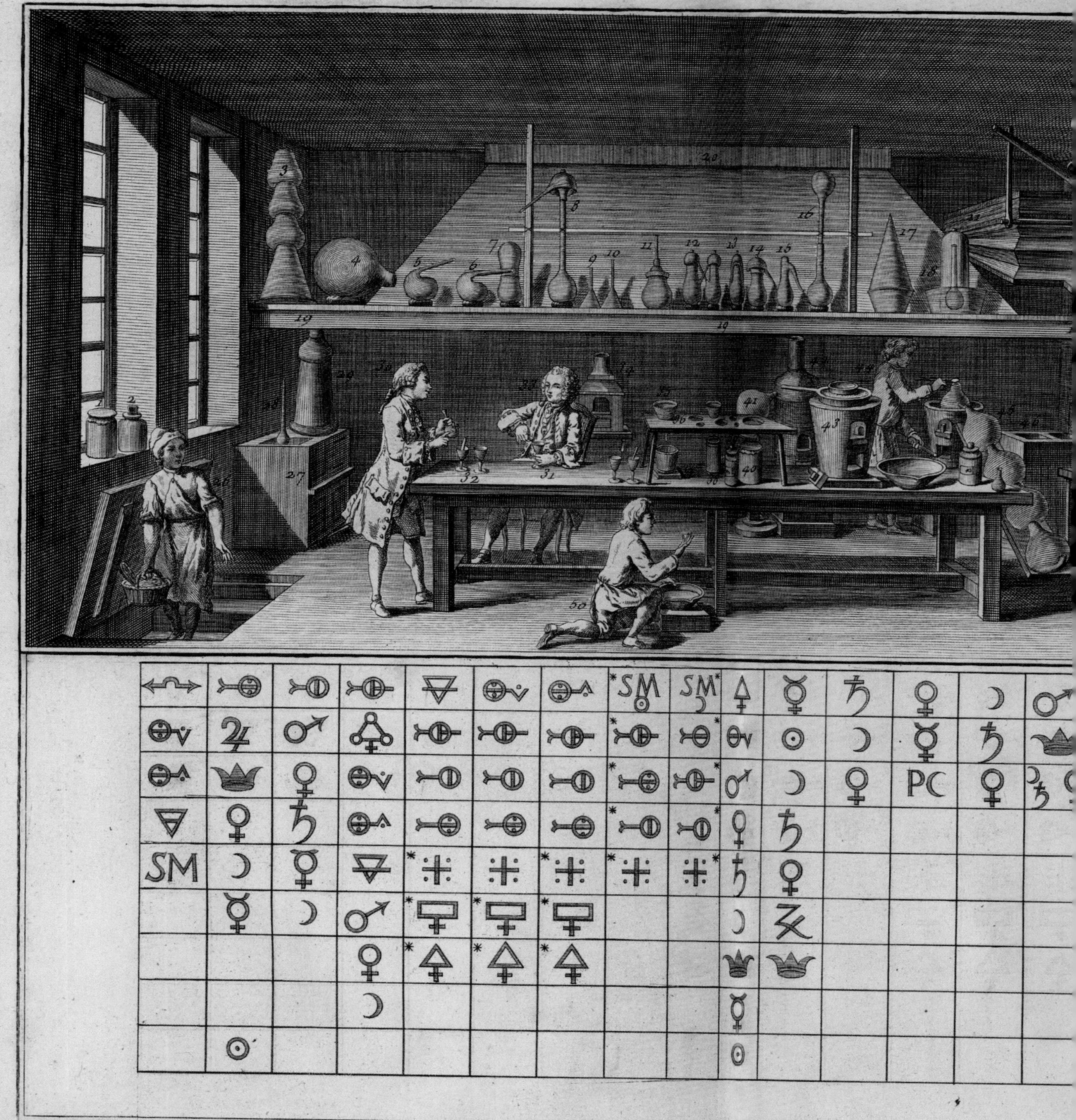

Goussier del.

Laboratoire et table des Raports

ILL. 62
A chemistry laboratory and affinity table of symbols representing different substances, a plate from *L'Encyclopédie, ou Dictionnaire raisonné des sciences, des arts et des métiers* (The Encyclopedia, or A Systematic Dictionary of the Sciences, Arts and Crafts), edited by Denis Diderot and Jean le Rond d'Alembert (Paris, 1751–72), plates vol. 3 (1763)
Bibliothèque Mazarine, Paris. Object no. 2° 3442-24

alkalis, the metals associated with the planets and the compounds are arranged in columns according to how they react with one another. This traditional nomenclature, based on the four ancient elements – earth, air, fire and water – had already been questioned, and the *Traité élémentaire de chimie*, published by the pioneering chemist Antoine Lavoisier in 1789, would permanently discredit it by generalising the names of the simple elements (such as sulphur) and distinguishing the acids by the suffixes '-ic' or '-ous' (sulphuric, sulphurous) and the salts by the suffixes '-ure' or '-ate' (sulphates) (Ill. 63).[7]

Below the chimney hood, on the furnace – a most important piece of laboratory equipment – are several models of apparatus made of earthenware or iron, for use in distillation, the reduction of salts or enamelling. Another essential feature, clean water, is supplied by a small tank on a stand, covered with wickerwork and

ILL. 63
Portrait of Antoine Laurent Lavoisier and Marie Anne Lavoisier (Marie Anne Pierrette Paulze), by Jacques Louis David, 1788
The Metropolitan Museum of Art, New York, Purchase, Mr. and Mrs. Charles Wrightsman Gift, in honor of Everett Fahy, 1977. Object no. 1977.10

ILL. 64
Alembic used for chemical distillation, from Antoine Laurent Lavoisier's laboratory, second half of the eighteenth century
Musée des Arts et Métiers – CNAM, Paris. Object no. 36645-0000-

furnished with a tap. While the earthenware and glass vessels are arranged on the shelves, the furnace implements hang beneath the hood. The scales, a precision instrument essential to the chemist, are carefully stored. Two tables or work surfaces are located on either side of the fireplace. The arrangement is completed by two mortars – one brass, the other marble – and a trestle table on which the long-necked flasks are drying upside down.

Despite the quality of the execution of the small parts, all created on the same scale of 1:8, it is a long way from the modern laboratory of the day: none of the new electrostatic machines, Leyden jars or apparatus for the study of gases that could be found at the time in Lavoisier's laboratory in the Paris Arsenal are to be seen (Ill. 64). What we see is an old-fashioned vision of the science of chemistry. It remains nevertheless an interesting model for studying historical instruments and is suggestive of the eighteenth-century fascination with chemistry for revealing nature's secrets.

NOTES

1 Hilaire-Perez 2008, pp. 36–41.
2 Saule and Arminjon 2010–11, p. 185.
3 Anon. 1785, pp. 57–58.
4 Corcy and Demeulenaere-Douyère 2022, pp. 131–45.
5 Diderot and d'Alembert 1751–65, text vol. 3 (1753), pp. 408–37.
6 Diderot 1762–72, plates vol. 3 (1763).
7 Bensaude-Vincent and Stengers 1992, pp. 116–21.

15. Mapping La Pérouse's Expedition: Louis XVI and the Pacific

JULIE GAREL-GRISLIN

From 1660, England basked in the glow of the discoveries made by the Royal Society and France could no longer ignore the crucial importance – as much military and strategic as diplomatic and political – of scientific research. In 1666, following the advice of his minister Jean-Baptiste Colbert, Louis XIV founded the Royal Academy of Sciences.[1]

Mastery of the seas was an essential asset in the struggle between the great powers of Europe. Exploration of the world opened up new trade routes and extended a country's political influence at international level. As a result, scientific activities gathered pace and enhanced the prestige of the nations that sponsored them.[2] Louis XIV launched French expeditions around the world 'by order of the king'; these missions had several objectives that were at once scientific, military and commercial.[3] Louis XV intensified this policy of state support, which culminated in the expedition of Jean-François de Galaup, comte de La Pérouse, at the behest of Louis XVI between 1785 and 1788.

The Treaties of Paris and Versailles, signed on 3 September 1783, marked the end of the American War of Independence (1775–83), which had opposed France and Britain for several years. However, rivalry continued between the two powers and was played out in competing scientific expeditions, a field that had opened up after the Seven Years War ended in 1763.[4] The eighteenth century witnessed a shift as the theatre of military operations moved to the seas, emphasising the importance of a powerful navy.[5] British maritime supremacy caused a certain amount of anxiety in French diplomatic circles, which were well aware of the significant political and economic stakes of the voyages undertaken by the Admiralty since the 1760s.[6] James Cook's three voyages between 1768 and 1779 resulted in important developments in the field of cartography, particularly with the accurate mapping of many Pacific coastlines and the dispelling of geographical myths, such as the existence of Terra Australis, a long-hypothesised continent in the southern hemisphere.[7] In doing so, Cook's voyages considerably reinforced Britain's strategic and commercial positions. They also fired the enthusiasm of Europe and of Louis XVI, who held a deep admiration for the English navigator and read the accounts of his expeditions (Ill. 65).

In France, the success of the voyage of Louis-Antoine de Bougainville (1766–69) and the European discovery of Tahiti were beginning to recede into memory, and only a large-scale venture in the wake of Cook's achievements could establish the maritime power of the kingdom (Ill. 66). Louis XVI gave his full support to a

ILL. 65
View of Tahiti from James Cook's second voyage, a plate from *Voyage dans l'hémisphère austral, et autour du monde ...* (A Voyage Towards the South Pole, and Round the World ...), translated into French after the original written by James Cook, Tobias Furneaux and Johann Reinhold Forster (Paris, 1778), vol. 1.
Bibliothèque municipale de Versailles. Object no. Res in-4 I 112 a

project that not only matched the spirit of the Age of Enlightenment but also set France on a course to open up new trade routes and provide opportunities to take possession of lands as yet unknown to Europeans, where ports, trading posts and colonies could be established. These investments were all the more crucial because they would compensate for the territorial losses suffered by France in America as a result of its defeat in 1763.

This exceptional voyage was sketched out in the autumn of 1783. It initially involved setting up trading posts in Alaska and developing a new commercial route with China intended for the export of furs from North America. Driven by Louis XVI and his advisers – in particular Charles-Pierre Claret, comte de Fleurieu, the director of ports and arsenals – the expedition was expanded to include political and scientific missions. Genuinely interested in geography and science, the king took an active part in the preparations. In the summer of 1785, he summoned the naval officer the comte de La Pérouse and presented him with the *Instructions*, a long text listing, page by page, the geographical, ethnological, scientific and economic objectives of the voyage (Ill. 67).[8] Thus, many savants embarked alongside sailors: astronomers, engineers, naturalists and even physicians charged with scrupulously following the encyclopedic programme of instructions set for them. Despite the rivalry between the major nations of Europe, there was also some cooperation. While the French geographers and astronomers prepared for their voyage drawing on the methods developed by Cook, Sir Joseph Banks succeeded in persuading the Royal Society to entrust France with two dip circles, a type of compass, that Cook had used on his expedition. The French were willing

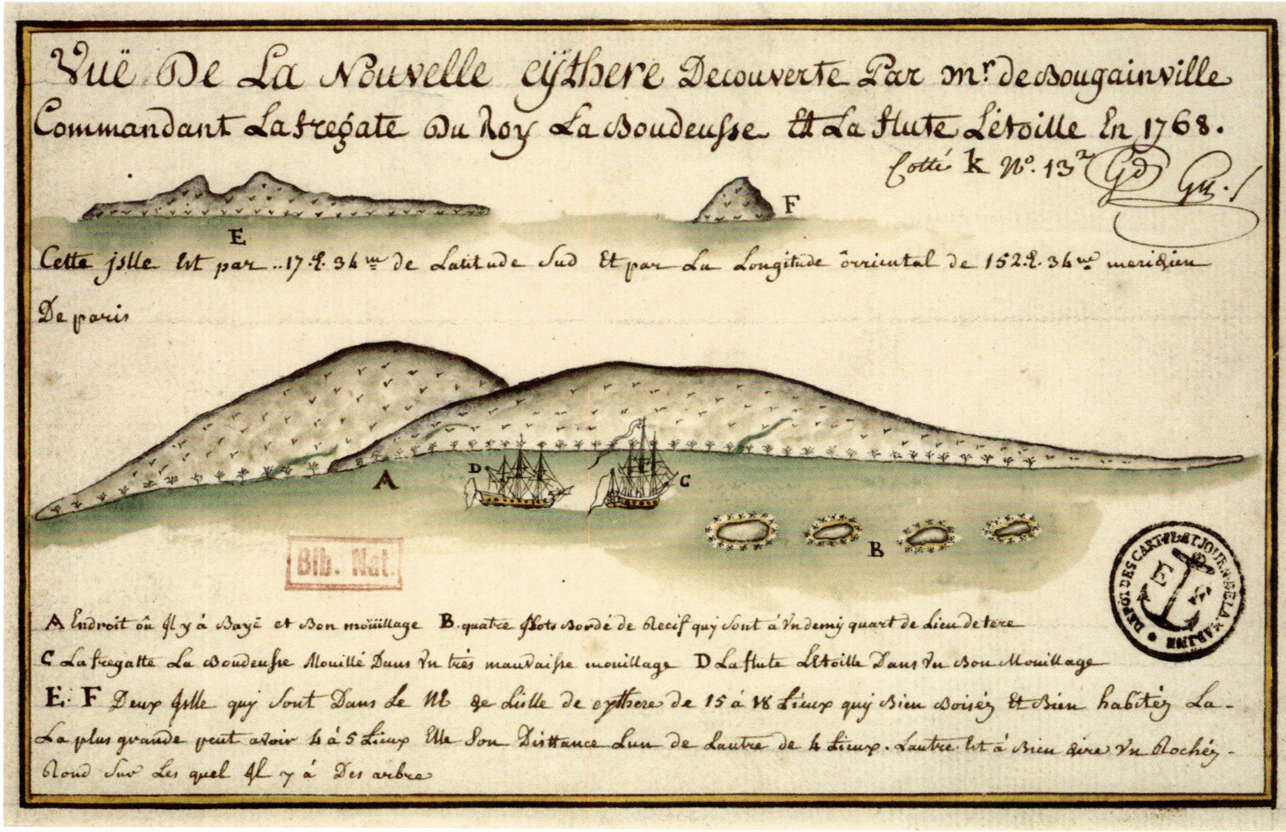

to borrow from their illustrious predecessor and indeed hoped to surpass his achievements thanks to their minutely detailed plans. As their cabins were being filled with books and small instruments, the holds of the two ships chartered for this mission, the *Astrolabe* and the *Boussole*, were being loaded with scientific equipment.[9] On 1 August 1785 the hour of departure struck. La Pérouse, with Paul Fleuriot de Langle as second in command, set out to conquer the seas, at the head of the greatest exploratory expedition of the age.

In order to observe, survey and trade, it was necessary to identify one's position. In addition to their navigational and measuring instruments, the sailors could rely on the maps prepared by the king's leading geographer, Jean-Nicolas Buache de la Neuville (Ill. 68).[10] These included a map of the Pacific Ocean on three large sheets of paper, showing the entire itinerary that La Pérouse was to follow, a route that wove in and out of those followed by the navigators who had crossed the Pacific between the sixteenth and eighteenth centuries (Álvaro de Mendaña y Neira, Pedro Fernandes de Queirós, Abel Tasman, Samuel Wallis, Bougainville and Cook).

This map captured the current state of knowledge, including the geographical grey areas that remained in 1785. The known coasts were drawn in solid lines, while dotted lines were used for those that could not yet be traced with certainty. For instance, the gap in the north-west coast of America between Cape Mendocino and the port of Bucareli only gave way to solid lines for coasts previously identified by Cook or Queirós.[11]

ILL. 66
Vuë de La Nouvelle Cÿthère découverte par Mr de Bougainville …
(View of New Cythera [Tahiti] Discovered by Mr de Bougainville …), 1768
Bibliothèque nationale de France, département des Cartes et plans, Paris. Object no. GE SH 18 PF 176 DIV 7 P 1 RES

ILL. 67
Louis XVI donnant ses Instructions *à La Pérouse, le 26 juin 1785* (Louis XVI Giving his *Instructions* to La Pérouse, 26 June 1785), by Nicolas-André Monsiau, 1817
Musée national des châteaux de Versailles et de Trianon. Object no. MV 220

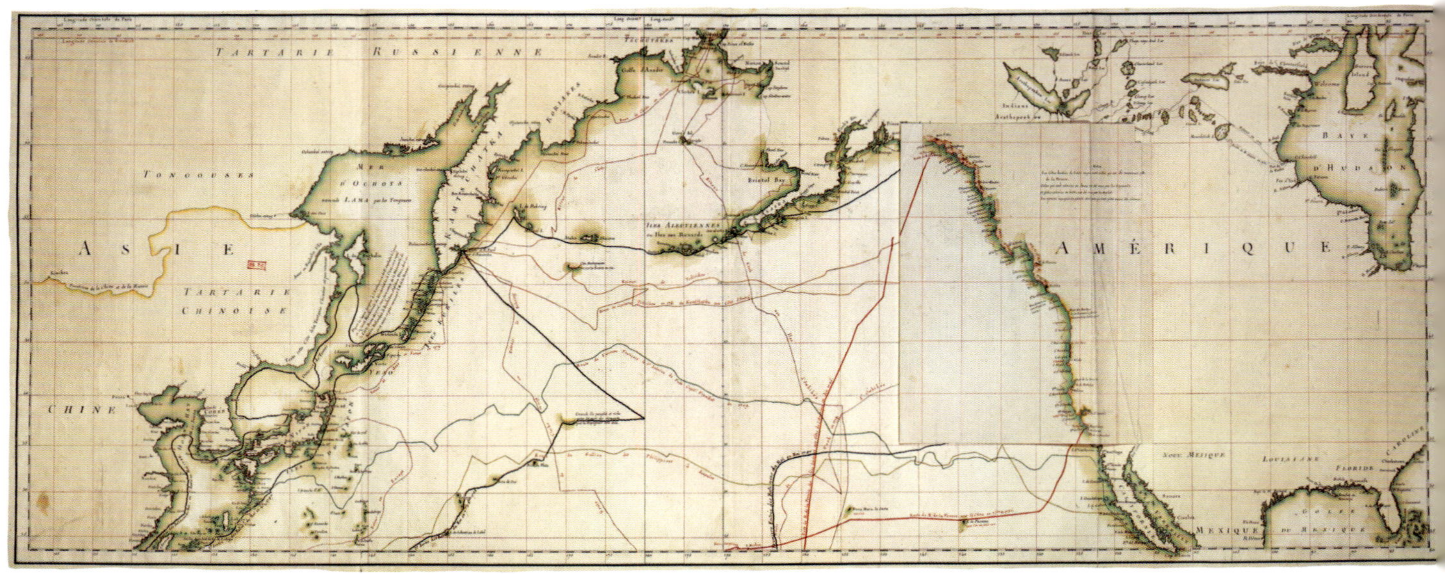
TARTARIE RUSSIENNE
TONGOUSES
MER D'OCHOTS
ASIE
TARTARIE CHINOISE
CHINE
ISLES ALEUTIENNES
Bristol Bay
AMÉRIQUE
BAYE D'HUDSON
NOUV. MEXIQUE
LOUISIANE
FLORIDE
MEXIQUE
GOLFE DU MEXIQUE

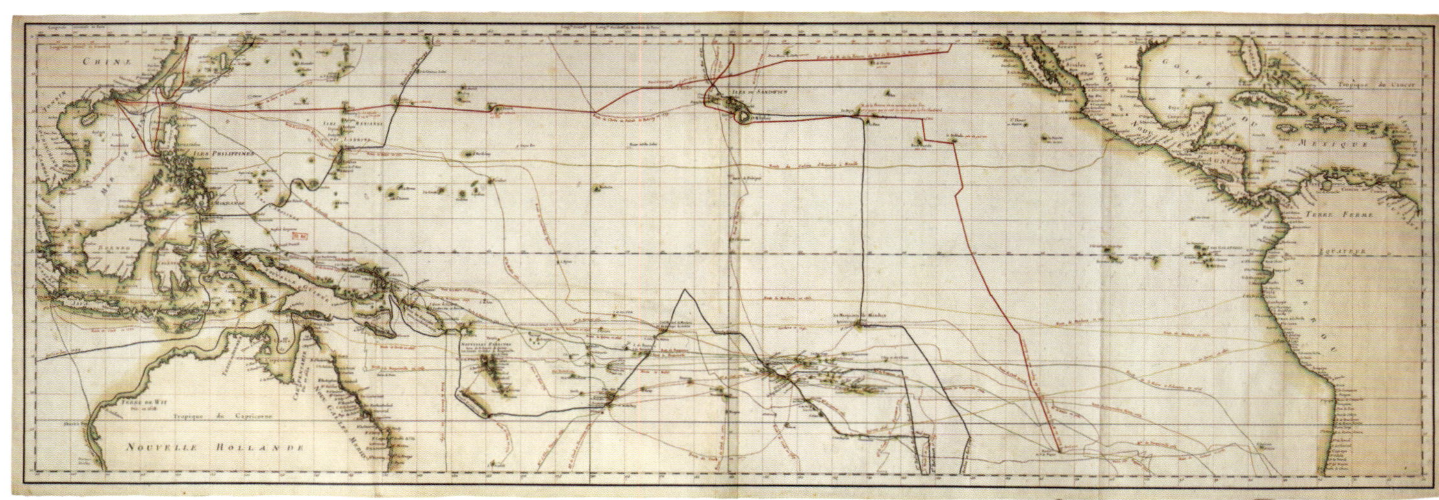
CHINE
ISLES PHILIPPINES
ISLES DE SANDWICH
TERRE DE WIT
Tropique du Capricorne
NOUVELLE HOLLANDE

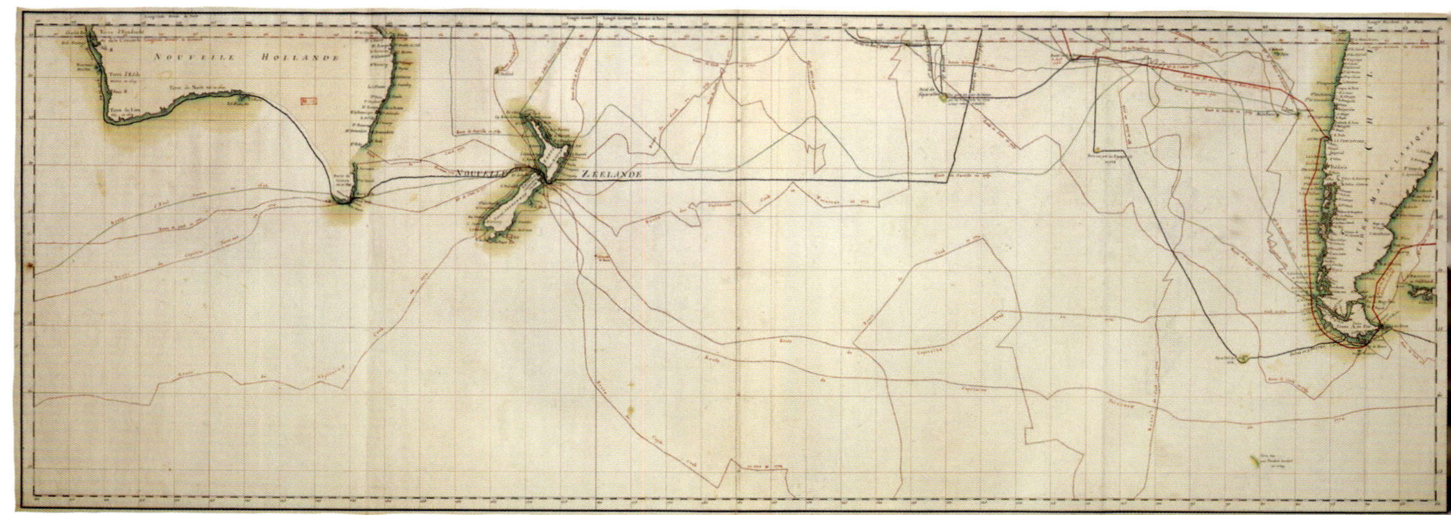
NOUVELLE HOLLANDE
NOUVELLE ZEELANDE

ILL. 68
Map owned by Louis XVI showing European voyages across the Pacific, including the itinerary and, marked in red, the actual route of La Pérouse, by Jean-Nicolas Buache de la Neuville, 1785
Bibliothèque nationale de France, département des Cartes et plans, Paris. Object no. GE SH 18 PF 174 P 1/2 RES

Buache made five copies of the map, one of which, lined with blue silk, was intended for the king. Louis XVI insisted on being informed in detail of the development of the expedition, and he scrupulously annotated his personal copy as La Pérouse progressed. The map was completed by a flap, on which the results of the reconnaissance of the Pacific coast of North America were transcribed. Adjoining the curly black line marking the expedition's itinerary is the route actually followed by La Pérouse, in red ink. This line ended tragically when the two ships were wrecked on the reefs of Vanikoro in 1788.

NOTES

1 Hahn 1971.
2 Pelletier 1998, pp. 112–27.
3 Schaffer 2005, pp. 791–815.
4 Brosse 1983; Baugh 2021.
5 Baugh 1994.
6 Kemp and Dear 2006.
7 Barnett 2018; Carter 1995, pp. 245–60; Frame and Walker 2018; Regard 2009; Williams 2010.
8 Kury 1998, pp. 65–91.
9 On the preparations and the expedition, see Gaziello 1984, pp. 35–166.
10 Broc 1971.
11 Gaziello 1984, pp. 83ff.

16. The Flight of the Montgolfière at Versailles

ANNA FERRARI

The excitement is almost palpable, the clamour audible, in this print representing Etienne Montgolfier's hot-air balloon demonstration at the Palace of Versailles on 19 September 1783 (Ill. 69). The flight was a momentous event and the culmination of a feverish summer of competing balloon ascents that started in June 1783 when brothers Etienne and Joseph Montgolfier, paper manufacturers from southern France, first flew a paper balloon filled with hot air that became known as a 'Montgolfière'. Within weeks, a parallel invention, the hydrogen balloon conceived by physics professor Jacques Charles and brothers Anne-Jean and Marie-Noël Robert, was flown from the Champ de Mars in Paris, drawing vast crowds. In September, Etienne Montgolfier was invited to showcase his hot-air balloon in the courtyard of the Palace of Versailles. A master stroke of propaganda, the ascent was staged against the backdrop of the sumptuous palace, firmly claiming the invention as French and associating it with the monarchy.

In the foreground, the print depicts a dense and colourful crowd of more than 130,000 people who had poured in from Paris, gathering around the launch platform, with figures jostling at the windows and even on the rooftops. Louis XVI and Marie-Antoinette watched from the palace, the king from the balcony of his room, the queen and court from under the canopy on the left-hand wing. The flight was both spectacle and experiment: a sheep, a rooster and a duck – the first living creatures to fly in a balloon – were placed in a wicker cage suspended from the balloon to demonstrate its safety. The anonymous printmaker chose to represent a moment of heightened drama when the balloon, having taken off, was whipped by the wind and tilted, losing some of its hot air and threatening to collapse.[1] Fortunately, however, the balloon righted itself and continued its flight for about 3 kilometres, to the satisfaction of both the king and queen who invited Etienne to their apartments later that day. The text below the image highlights the balloon's regal blue and gold colours, and the king's cipher of intertwined Ls, as well as a precise record of its size (equivalent to 18.5 metres high and 13.3 metres in diameter), weight and carrying capacity, underlining the invention's scientific character.

The ascents sparked a balloon craze and a flurry of balloon-themed textiles, jewellery, furniture and decorative objects (Ill. 70), as well as thousands of single-sheet prints such as this one opposite published by Le Noir in Paris, which disseminated across Europe the incredible new spectacle of flight.[2] Benjamin Franklin, the American polymath, politician and diplomat who was in Paris for

ILL. 69
Print showing the demonstration of the Montgolfière balloon at Versailles on 19 September 1783, published by Le Noir, 1783
Science Museum Group, London. Object no. 1979-527/3

ILL. 70
Varicoloured gold oval snuffbox, with on the lid the motif of a trophy composed of a balloon and scientific instruments, 1783–95
Victoria and Albert Museum, London. Object no. 172-1878

ILL. 71
Montgolfier in the Clouds: Constructing of Air Balloons for the Grand Monarque, published by Samuel William Fores, 2 March 1784
Science Museum Group, London. Object no. 1950-303/2

peace negotiations, witnessed the flight of Charles's hydrogen balloon, reporting that balloons were the subject of all conversations in Paris in autumn 1783.[3]

Balloons emerged in the context of pneumatic research and rapidly changing knowledge about the chemistry of air. Discoveries in the 1760s and 1770s revealed that air was composed of different gases with different weights and chemical properties. In England, Henry Cavendish published his discovery of 'inflammable air', a gas much lighter than air (which became known as hydrogen), in 1766, while Joseph Priestley published *Experiments and Observations on Different Kinds of Air* between 1774 and 1786. This research was swiftly translated and relayed in French academic circles, spurring Antoine Lavoisier's experiments with air and the gas he named oxygen in 1778. The topic captured the imagination of the Montgolfier brothers who acquired a copy of Priestley's *Experiments and Observations*, furthering their understanding of the properties of hot air. In November 1783, when Joseph spoke at the Académie des Sciences, Belles Lettres et Arts de Lyon (Academy of Sciences, Literature and Arts of Lyon) he explained: 'We have imagined [...] to enclose in a light vessel a fluid specifically less heavy than atmospheric air, in order to benefit from the broken equilibrium between these two fluids, to raise in the air masses proportionate to the volume of the rising vessel.'[4]

The story of the Montgolfière was also entwined with France's foreign policy following its humiliating defeat against the British in the Seven Years War (1756–63) and its subsequent engagement in the American War of Independence (1775–83).[5] In a retrospective account of the origins of the hot-air balloon, Joseph Montgolfier claimed the idea occurred to him in November 1782 while looking at a print representing the siege of Gibraltar. At the time, France was engaged in this war, which extended to Europe. During the hostilities, France and Spain blockaded the strategic fortress of Gibraltar that had been under British control

MONTGOLFIER IN THE CLOUDS
O by gar! dis be de grande invention. — Dis will immortalize my King, my Country, and myself;
We will declare de War against our ennemi; we will make des English quake, by gar:
We will inspect their Camp, we will intercept their Fleet, and we will set fire to their Dock-yards:
And by gar, we will take de Gibraltar in de air balloon, and when we have
Conquered d' Eenglish, den we conquer d' other Countrie, and make them all Colonie
to de Grand Monarque.
CONSTRUCTING OF AIR BALLOONS FOR THE GRAND MONARQUE
FOURTH SKETCH
a Companion to this in a few days
Published as the act directs March 2 1784 by S. Fores No. 3 Piccadilly

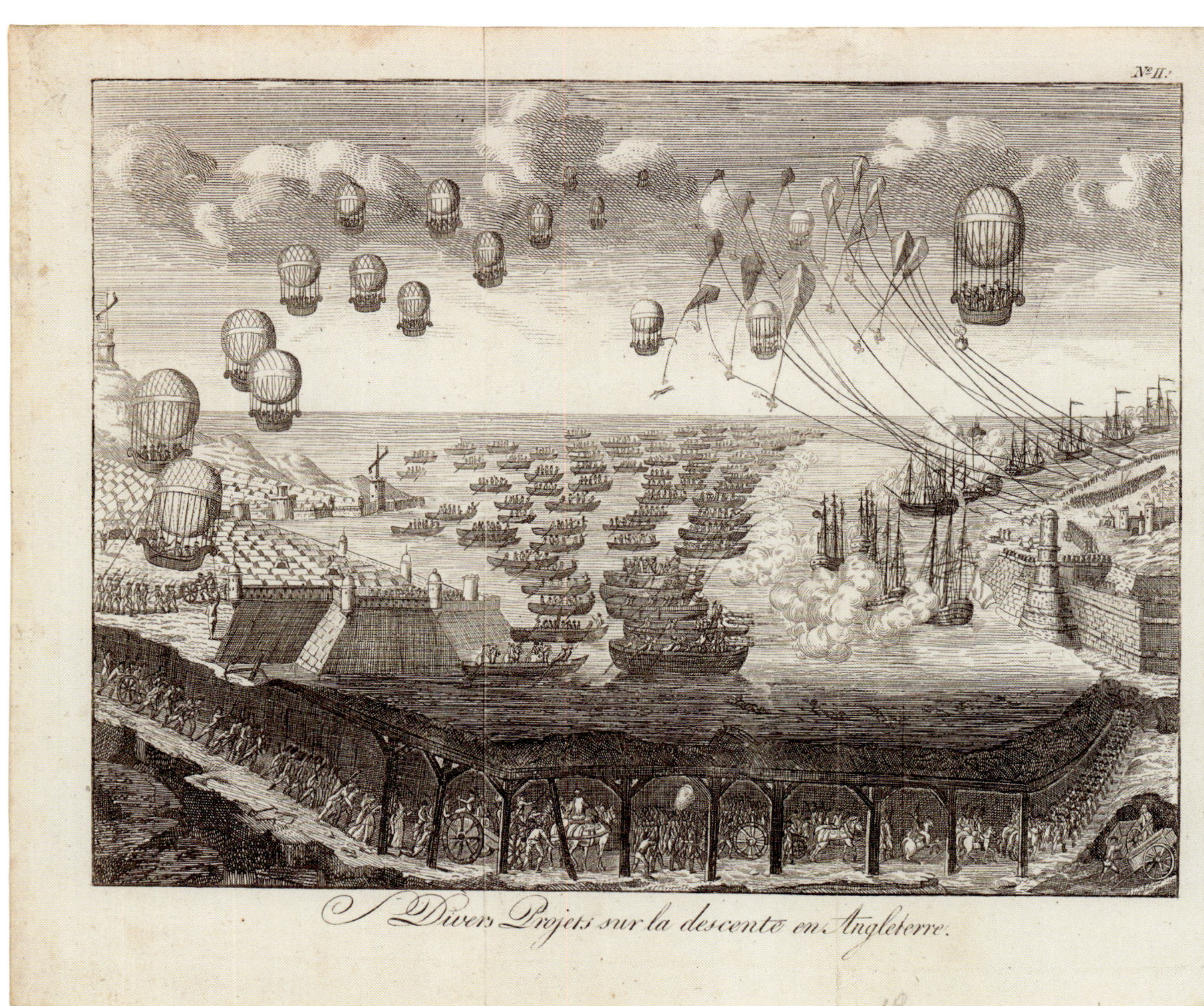

ILL. 72
Divers Projets sur la descente en Angleterre (Various Plans for the Invasion of England), mid-1790s to 1800s
Science Museum Group, London. Object no. 1984-483

since 1713. The sight of smoke rising up a chimney allegedly prompted Joseph to imagine a way to end the siege by harnessing the power of hot air to reach Gibraltar by air, rather than land or sea.[6]

By the time the Montgolfière was ready, the American War of Independence had ended. The signing of the peace treaties recognising the independence of the American colonies took place in early September 1783 in Paris and Versailles. France emerged with little to show except worsening debts. However, Montgolfier's demonstration was an opportunity to promote France as an innovative kingdom while international diplomats were still present.[7] Against this political backdrop, the potential military advantages of the invention were often highlighted: resupplying a besieged city, carrying out reconnaissance missions and transporting troops.[8]

Royal support for developing balloons continued with the first human ascent occurring just a few weeks later on 21 November 1783 at the Château de la Muette, another royal residence near Versailles, with the marquis François Laurent d'Arlandes and the chemist Jean-François Pilâtre de Rozier becoming the first aeronauts. Following these successes, in January 1784 the Minister of Royal Finances recommended the award of 8,000 livres to Etienne Montgolfier to allow him to build a balloon to cross the Channel.[9] The British ridiculed such French ambitions in a print published later that year, showing one of the Montgolfier brothers perched in the clouds blowing balloon-like soap bubbles with a pipe (Ill. 71). The text below, spelt to suggest a crude French accent, mocks Montgolfier's pride in his 'grande invention' and grandiose plans to 'declare de War against our ennemi', 'make des English quake' and 'take de Gibraltar in de air balloon'.

By 1785 balloon mania had begun to wane, coinciding with Pilâtre de Rozier's death in June during his attempt to cross the Channel in a balloon commissioned by the royal administration. Yet the dream of deploying balloons against the British survived the French Revolution and a print from the mid-1790s to early 1800s presents a fantastical project for an air, sea and underground invasion across the Channel (Ill. 72).

NOTES

1 From Etienne's letter to his wife Adélaïde, quoted in Gillispie 1983, p. 42.
2 The Science Museum holds a collection of aeronautica, including prints, books and decorative objects relating to ballooning, bequeathed by Winifred Penn-Gaskell in 1949.
3 Franklin 1904, 16 September 1783.
4 Thébaud-Sorger 2009, translated from the French by the present author.
5 Kim 2013, p. 853.
6 Gillispie 1983, p. 16.
7 Thébaud-Sorger 2009.
8 Montgolfier 1784, p. 9.
9 Hirschauer 1917, pp. 59–60.

Bibliography

Ancelin, Justine, *Science, académisme et sociabilité savante : édition critique et étude du Journal de la vie privée de Jean-Dominique Cassini (1710–1712)* (Paris, 2011)

Anon., 'De la figure de la lune et de son usage', in *Connoissance des Temps pour l'Année 1751* (Paris, 1750), pp. 196–98

Anon., *Gazette de France*, March 1754a, pp. 105–6

Anon., *Mercure de France*, May 1754b, pp. 152–56

Anon., 'Collection des Modèles des Arts & Métiers, & Manufactures', in *Almanach du Palais-Royal* (Paris, 1785), pp. 57–58

Bancel, Nicolas et al., *Histoire globale de la France coloniale* (Paris, 2022)

Baridon, Michel, *A History of the Gardens of Versailles*, trans. Adrienne Mason (Philadelphia, 2008)

Barnett, James K., *Captain Cook's Final Voyage: The Untold Story from the Journals of James Burney and Henry Roberts* (Washington, 2018)

Baugh, Daniel, 'Maritime strength and Atlantic commerce: the uses of "a Grand Marine Empire"', in Laurence Stone (ed.), *An Imperial State at War: Britain from 1689 to 1815* (London, 1994)

Baugh, Daniel, *The Global Seven Years War, 1754–1763: Britain and France in a Great Power Contest* (London, 2021)

Bensaude-Vincent, Bernadette and Christine Blondel (eds), *Science and Spectacle in the European Enlightenment* (Aldershot and Burlington VT, 2008)

Bensaude-Vincent, Bernadette and Isabelle Stengers, *Histoire de la chimie* (Paris, 1992)

Beretta, Marco and Paolo Brenni, *The Arsenal of Eighteenth-Century Chemistry: The Laboratories of Antoine Laurent Lavoisier (1743–1794)* (Leiden, 2022)

Bertucci, Paola, *Artisanal Enlightenment. Science and the Mechanical Arts in Old Regime France* (New Haven and London, 2017)

Bertucci, Paola, *In the Land of Marvels. Science, Fabricated Realities, and Industrial Espionage in the Age of the Grand Tour* (Baltimore, 2023)

Besterman, Theodore (ed.), *Correspondence de Voltaire*, 107 vols (Geneva, 1953–65)

Besterman, Theodore (ed.), *Correspondence and Related Documents, Œuvres complètes de Voltaire*, vols 85–135 (Oxford, 1968–77)

Bimbenet-Privat, Michèle, Jannic Durand and Frédéric Dassas, *Décors, mobilier et objets d'art du musée du Louvre, de Louis XIV à Marie-Antoinette* (Paris, 2014)

Boutard, François, *Ode Latine sur Marly, traduite en françois. Au roy*, trans. into French by Charles Perrault (Paris, 1697)

Brandstetter, Thomas, 'La machine de Marly: un spectacle technologique', *Bulletin du Centre de recherche du château de Versailles*, 5 (2012), http://journals.openedition.org/crcv/11907 (accessed 8 May 2024)

Broc, Numa, 'Un géographe dans son siècle, Philippe Buache (1700–1773)', *Dix-huitième Siècle*, vol. 3 (1971), pp. 223–35

Brockliss, Laurence and Colin Jones, *The Medical World of Early Modern France* (Oxford, 1997)

Brosse, Jacques, *Les Tours du monde des explorateurs. Les grands voyages maritimes, 1764–1843* (Paris, 1983)

Burke, Peter, *Fabrication of Louis XIV* (New Haven and London, 1992)

Carlier, Yves and Hélène Delalex (eds). *Louis XV. Passions d'un roi* (Château de Versailles, exh. cat., 2022–23)

Carré, Anne-Laure (ed.), *Top Modèles. Une leçon princière au XVIIIe siècle* (Paris, 2020)

Carter, Harold B., 'The Royal Society and the voyage of HMS Endeavour 1768–1771', *Notes and Records of the Royal Society of London*, vol. 49, no.2 (1995), pp. 245–60

Cassini, Jean-Dominique, 'Avertissement touchant l'observation de l'Éclipse de Lune, qui doit arriver la nuit du 28 Juillet prochain', in *Mémoires de Mathématique et de Physique Tirez des Registres l'Académie Royale des Sciences* (Paris, 1692a), pp. 111–12

Cassini, Jean-Dominique, 'Observation de l'Éclipse de Lune du 28 Juillet dernier', in *Mémoires de Mathématique et de Physique Tirez des Registres l'Académie Royale des Sciences* (Paris, 1692b), pp. 129–35

Castelluccio, Stéphane, 'Louis XIV, le Siam et la Chine: Séduire et être séduit', *Extrême-Orient Extrême-Occident*, vol. 43 (2019), pp. 25–44

Castelluccio, Stéphane, 'Le Palais de Trianon dans la vie de Cour', *Versalia*, no. 27 (2024), pp. 109–30

Corcy, Marie-Sophie and Christiane Demeulenaere-Douyère, 'Une revendication inopinée de paternité pour les modèles de Madame de Genlis: le mystérieux "artiste mécanicien" Mathieu Laveron', in Anne-Laure Carré and Martine Reid (eds), *Des modèles à l'étude. L'éducation des princes au XVIIIe siècle* (Rouen, 2022), pp. 131–45

Dassas, Frédéric and Marc Voisot, 'Pendule de la Création du Monde', in Carlier and Delalex 2022–23, p. 250.

Delalex, Hélène, 'Discovery of the celestial and terrestrial worlds: astronomy and great expeditions', in Hélène Delalex and Bertrand Rondot (eds), *Versailles and the World*, trans. Charles Penwarden (Louvre Abu Dhabi, exh. cat., 2022), pp. 163–69

Despoix, Philippe, 'Mesure du monde et représentation européenne au XVIIIe siècle: Le programme britannique de détermination de la longitude en mer / Measurement of the world and European representation in the 18th century: the British programme to measure longitudes at sea', *Revue d'histoire des sciences*, vol. 53, no. 2 (2000), pp. 205–33

Diderot, Denis (ed.), *Recueil de planches sur les sciences, les arts libéraux, et les arts méchaniques avec leur explication*, 11 vols (Paris, 1762–72)

Diderot, Denis, *Rameau's Nephew and D'Alembert's Dream*, trans. Leonard Tancock (Harmondsworth, 1966)

Diderot, Denis, *Diderot on Art – II The Salon of 1767*, ed. and trans. John Goodman, introduction by Thomas Crow (New Haven and London, 1995)

Diderot, Denis and Jean le Rond d'Alembert (eds), *L'Encyclopédie, ou Dictionnaire raisonné des sciences, des arts et des métiers*, 17 vols (Paris, 1751–65)

Donneau de Visé, Jean, *Suite du voyage des ambassadeurs de Siam en France … novembre 1686, seconde partie* (Paris, 1686)

Dunn, Richard and Rebekah Higgitt, *Finding Longitude: How Clocks and Stars Helped Solve the Longitude Problem* (London, 2014)

Du Tertre, Jean-Baptiste, 'Chapitre II. Des plantes qui portent des fruits. De l'Ananas, & des Karatas à fruits', in Jean-Baptiste Du Tertre, *Histoire générale des Antilles habitées par les Français*, 4 vols (Paris, 1667–71)

Easterby-Smith, Sarah, *Cultivating Commerce: Cultures of Botany in Britain and France, 1760–1815* (Cambridge, 2018)

Etablissement public du château, du musée et du domaine national de Versailles, secteur éducatif, *Dossier pédagogique: La machine de Marly* (n.d.), online edition, https://www.chateauversailles.fr/sites/default/files/presse/documents/ressource-pedagogique_dossier-machine_marly.pdf (accessed 9 May 2024)

Fara, Patricia, *Pandora's Breeches: Women, Science and Power in the Enlightenment* (London, 2004)

Fernández de Oviedo y Valdés, Gonzalo, *Historia general y natural de las Indias* (part 1) (Seville, 1535)

Frame, William and Laura Walker, *James Cook: The Voyages* (London, 2018)

Franklin, Benjamin, *The Works of Benjamin Franklin, Vol. X Letters and Misc. Writings 1782–1784* (New York, 1904), https://oll.libertyfund.org/title/franklin-the-works-of-benjamin-franklin-vol-x-letters-and-misc-writings-1782-1784 (accessed 3 November 2023)

Gaziello, Catherine, *L'Expédition de Lapérouse (1785–1788): Réplique française aux voyages de Cook* (Paris, 1984)

Gelbart, Nina Rattner, 'The monarchy's midwife who left no memoirs', *French Historical Studies*, vol. 19, no. 4, Special Issue: Biography (1996), pp. 997–1023

Gelbart, Nina Rattner, *The King's Midwife: A History and Mystery of Madame Du Coudray* (Berkeley, 1998)

Gelbart, Nina Rattner, *Minerva's French Sisters: Women of Science in Enlightenment France* (New Haven and London, 2021)

Gillispie, Charles Coulston, *The Montgolfier Brothers and the Invention of Aviation 1783–1784* (Princeton, 1983)

Gislén, Lars et al., 'Cassini's 1679 map of the moon and French Jesuit observations of the lunar eclipse of 11 December 1685', *Journal of Astronomical History and Heritage*, vol. 21, no. 2/3 (2018), pp. 211–25

Godfroy, Marion (ed.), *Kourou, 1763 – Le dernier rêve de l'Amérique française* (Paris, 2011)

Grimbergen, Kees, 'Huygens and the advancement of time measurements', in Karen Fletcher (ed.), *Titan – From Discovery to Encounter* (Noordwijk, 2004), pp. 91–102

Guerrini, Anita, *The Courtiers' Anatomists: Animals and Humans in Louis XIV's Paris* (Chicago, 2015)

Hahn, Roger, *The Anatomy of a Scientific Institution: The Paris Academy of Sciences, 1666–1803* (Berkeley, 1971)

Hahn, Roger, 'La Fondation de l'Académie royale des sciences', in Saule and Arminjon 2010–11, pp. 30–38

Haudrère, Philippe, *La Compagnie française des Indes au XVIIIe siècle (1719–1795)*, 2 vols (Paris, 2005)

Heilbron, John L., *Electricity in the 17th and 18th Centuries. A Study of Early Modern Physics* (1979) (Berkeley, 2022)

Hilaire-Perez, Liliane, 'Technology, curiosity and utility in France and England in the eighteenth century', in Bensaude-Vincent and Blondel 2008, pp. 25–42

Hirschauer, Charles, *Les Premières expériences aérostatiques à Versailles (19 septembre 1783 – 23 juin 1784)* (Versailles, 1917)

Hoare, Michael, *Weighing Fire: European Lives in Eighteenth-century Literature and Science*, 2 vols (Oxford, 2022)

Hunter, Michael, *The Image of Restoration Science: The Frontispiece to Thomas Sprat's History of the Royal Society (1667)*, with a chapter on the instruments by Jim Bennett (Abingdon, 2017)

Jami, Catherine, 'Un empereur à l'observatoire: L'Histoire de la Chine au prisme d'une anecdote', *Écrire l'histoire*, no. 17 (2017), pp. 65–74

Jones, Colin, *The Great Nation: France from Louis XV to Napoleon* (London, 2003)

Jones, Mark, *Medals of the Sun King* (London, 1979)

Jullien, Vincent (ed.), *Le Calcul des longitudes. Un Enjeu pour les mathématiques, l'astronomie, la mesure du temps et la navigation* (Rennes, 2002)

Kemp, Peter and I.C.B. Dear, *The Oxford Companion to Ships and the Sea* (Oxford, 2006)

Kim, Mi Gyung, 'Invention as a social drama: from an ascending machine to the aerostatic globe', *Technology and Culture*, vol. 54, no. 4 (October 2013), pp. 853–87

Kim, Mi Gyung, *The Imagined Empire: Balloon Enlightenments in Revolutionary Europe* (Pittsburgh, 2016)

Kury, Lorelai, 'Les instructions de voyage dans les expéditions scientifiques françaises 1750–1830', *Revue d'histoire des sciences*, vol. 51 (1998), pp. 65–91

La Fontaine, Jean de, *Fables choisies, mises en vers par M. de La Fontaine*, Book 2 (Paris, 1682)

Lamy, Gabriela, '*Le Jardin d'Eden: Le Paradis terrestre renouvellé dans le jardin de la Reine à Trianon* de Pierre-Joseph Buc'hoz', *Bulletin du Centre de recherche du château de Versailles* (2010), https://doi.org/10.4000/crcv.10300 (accessed 9 May 2024)

Lamy, Gabriela, 'Trianon, jardin d'études et d'essais', in Saule and Arminjon 2010–11, pp. 140–51

Lamy, Gabriela, 'Les pérégrinations de l'ananas' in Anne de Thoisy-Dallem (ed.), *Parties de campagne. Jardins et champs dans la toile imprimée XVIIIe-XIXe siècles* (Musée de la toile de Jouy, exh. cat., 2011), pp. 107–15

Lamy, Gabriela, 'La Guyane évoquée à Trianon en 1783 à travers quelques représentations de plantes du *Jardin d'Eden, le paradis terrestre renouvellé dans le Jardin de la Reine à Trianon* de Pierre-Joseph Buc'hoz', in Pierre Rosenberg and Monique Mosser (eds), *De l'Art des jardins de papier: concevoir, projeter, représenter*, Conference proceedings of the XVèmes Rencontres Internationales du Salon du dessin (Paris, 2023a), pp. 67–82

Lamy, Gabriela, 'Men, plants and gardens between Mauritius and the Petit Trianon in the 18th century', *Studies in the History of Gardens & Designed Landscapes*, vol. 43, no. 1 (2023b), pp. 34–66

Le Lous, Maela, John Baxter and Krystel Nyangoh Timoh, 'Madame Angélique du Coudray: pioneer of medical simulation and unsung hero', *Journal of Gynecology Obstetrics and Human Reproduction*, vol. 52, no. 2 (2022), https://doi.org/10.1016/j.jogoh.2022.102529 (accessed 9 May 2024)

Le Roi, Joseph Adrien, *Notice historique sur le Potager du Roi à Versailles* (Versailles, 1847)

Léry, Jean de, *Histoire d'un voyage fait en la terre de Brésil* (La Rochelle, 1578)

Lilti, Antoine, *The World of the Salons: Sociability and Worldliness in Eighteenth-century Paris* (New York, 2015)

Loskoutoff, Yvan (ed.), *Les Médailles de Louis XIV et leur Livre* (Rouen and Le Havre, 2016)

Mahoney, Michael, 'Christiaan Huygens: the measurement of time and longitude at sea', in Henk J.M. Bos et al. (eds), *Studies on Christiaan Huygens. Invited Papers from the Symposium on the Life and Work of Christiaan Huygens, Amsterdam, 22–25 August 1979* (Lisse, 1980), pp. 234–70

Major, F.G., 'The longitude problem', *Quo Vadis: Evolution of Modern Navigation: The Rise of Quantum Techniques* (New York, 2014), pp. 113–29

Maral, Alexandre and Nicolas Milovanovic (eds) *Les Animaux du Roi* (Palace of Versailles, exh. cat., 2021–22)

McClellan III, James E. and François Regourd, 'The colonial machine: French science and colonization in the ancien regime', *Osiris*, vol. 15, Nature and Empire: Science and the Colonial Enterprise (2000), pp. 31–50

Montgolfier, Joseph, *Discours prononcé à l'Académie des Sciences, Belles Lettres et Arts de Lyon* (October 1783), (Paris 1784)

Morera, Raphaël, 'Amener les eaux: entre techniques, sciences et politique', in Saule and Arminjon 2010–11, pp. 87–93

Mukerji, Chandra, *Territorial Ambitions and the Gardens of Versailles* (Cambridge, 1997)

Pelletier, Monique, 'La France mesurée', *Mappemonde*, vol. 3 (1986), pp. 26–32

Pelletier, Monique, 'Cartographie et pouvoir sous les règnes de Louis XIV et Louis XV', in Danielle Lecqoq and Antoine Chambard (eds), *Terre à découvrir. Terres à parcourir. Exploration et connaissance du monde XIIe–XIXe siècles* (Paris, 1998), pp. 112–27

Plomp, Reinier, *Early French Pendulum Clocks, 1658–1700: Known as Pendules Religieuses* (Schiedam, 2009)

Plomp, Reinier, 'A longitude timekeeper by Isaac Thuret with the balance spring invented by Christiaan Huygens', n.d., http://www.antique-horology.org/_editorial/thuretplomp/thuretplomp.htm (accessed 20 November 2023)

Poirier, Jean-Pierre, *Histoire des femmes de science en France du Moyen Age à la Révolution* (Paris, 2002)

Pyenson, Lewis and Jean-François Gauvin (eds), *The Art of Teaching Physics: The Eighteenth-Century Demonstration Apparatus of Jean-Antoine Nollet* (Sillery, 2002)

Raynal, Guillaume Thomas, *Histoire philosophique et politique des établissements et du commerce des Européens dans les deux Indes*, 6 vols (La Haye, 1774)

Regard, Frédéric, *British Narratives of Exploration: Case Studies on the Self and Other* (London, 2009)

Regourd, François, 'Capitale savante, capitale coloniale: Sciences et savoirs coloniaux à Paris aux XVIIe et XVIIIe siècles', *Revue d'histoire moderne & contemporaine*, vol. 55, no. 2 (2008), pp. 121–51

Rookmaaker, Leendert Cornelis, 'Histoire du rhinocéros de Versailles (1770–1893)', *Revue d'histoire des sciences*, vol. 36, no. 3–4 (1983), pp. 307–18

Rouet, Marion, 'Cerisiers, pêchers et ananas: des cultures de pointe dans les potagers royaux en Île-de-France dans la seconde moitié du XVIIIe siècle', *In Situ*, no. 41 (2019), https://journals.openedition.org/insitu/25466?lang=fr (accessed 9 May 2024)

Sabatier, Gérard, *Versailles ou la disgrâce d'Apollon* (Rennes, 2017), online edition, https://books.openedition.org/pur/155490 (accessed 9 May 2024)

Sahut, Marie-Catherine and Nathalie Volle (eds), *Diderot et l'art de Boucher à David: Les Salons, 1759–1781* (Hôtel de la Monnaie, Paris, exh. cat., 1984–85)

Sarrazin, Jean-Yves, 'Belles et obsolètes: deux "machines" astronomiques', *Revue de la Bibliothèque nationale de France*, no. 14 (2003), pp. 46–47

Saule, Béatrix and Catherine Arminjon (eds), *Sciences et curiosités à la cour de Versailles* (Palace of Versailles, exh. cat., 2010–11)

Saule, Béatrix and Lucina Ward (eds), *Versailles: Treasures from the Palace* (National Gallery of Australia, Canberra, exh. cat., 2016–17)

Schaffer, Simon, 'L'inventaire de l'astronome. Le commerce d'instruments scientifiques au XVIIIe siècle (Angleterre-Chine-Pacifique)', *Annales. Histoire, Sciences Sociales*, vol. 60, no. 4 (2005), pp. 791–815

Schiebinger, Londa and Claudia Swan (eds), *Colonial Botany: Science, Commerce, and Politics in the Early Modern World* (Philadelphia, 2004)

Spary, Emma, *Utopia's Garden: French Natural History from Old Regime to Revolution* (Chicago, 2000)

Spary, Emma, *Feeding France: New Sciences of Food, 1760–1815* (Cambridge, 2014)

Stroup, Alice, *A Company of Scientists: Botany, Patronage, and Community at the Seventeenth-Century Parisian Royal Academy of Sciences* (Berkeley, 1990)

Stryienski, Casimir, *Mesdames de France: filles de Louis XV* (Paris, 1911)

Taub, Liba, 'What is a scientific instrument, now?', *Journal of the History of Collections*, vol. 31, no. 3 (2019), pp. 453–67

Thébaud-Sorger, Marie, 'Chapitre I. De l'invention à l'évènement', in Marie Thébaud-Sorger, *L'Aérostation au temps des Lumières* (Rennes, 2009), http://books.openedition.org/pur/100749 (accessed 30 October 2023)

Thompson, David, 'Historical timeline of clocks', *AHS: The Story of Time*, 2001–9, https://www.ahsoc.org/resources/historical-timeline/ (accessed 20 November 2023)

Turner, Anthony, *Mathematical Instruments in the Collection of the Bibliothèque Nationale de France* (Paris, 2018)

Voltaire, *Elémens de la philosophie de Neuton* (Amsterdam, 1738)

Werrett, Simon, *Fireworks: Pyrotechnic Arts and Sciences in European History* (Chicago, 2010)

Widemann, Thomas, 'Les observations astronomiques dans les résidences royales', in Saule and Arminjon 2010–11, pp. 220–24

Williams, Glyn, *Arctic Labyrinth: The Quest for the Northwest Passage* (Toronto, 2010)

Willms, Allan R., Petko M. Kitanov and William F. Langford, 'Huygens' clocks revisited', *Royal Society Open Science*, vol. 4, issue 9 (September 2017), https://royalsocietypublishing.org/doi/10.1098/rsos.170777 (accessed 20 November 2023)

Wolf, Charles, *Histoire de l'Observatoire de Paris de sa fondation à 1793* (Paris, 1902)

Zinsser, Judith P., *La Dame d'Esprit: A Biography of the Marquise du Châtelet* (New York, 2006)

Picture Credits

All images © The Board of Trustees of the Science Museum unless stated otherwise.

Front endpaper and pp. 81, 88, 109: © Bibliothèque municipale de Versailles
Back endpaper and p. 58: © Sébastien Giles
Frontispiece and pp. 18, 28, 43, 45, 47, 52, 75, 80, 110, 112: Bibliothèque nationale de France
p. 8: Image courtesy The Metropolitan Museum of Art, New York
pp. 11, 76, 99, 111: © Château de Versailles, Dist. GrandPalaisRmn / Christophe Fouin
p. 14: Public domain. Photo: © Christian Devleeschauwer
pp. 15, 25, 38, 55: © GrandPalaisRmn (Château de Versailles) / Gérard Blot
pp. 17, 56: Royal Collection Trust / © His Majesty King Charles III 2024
p. 19: © GrandPalaisRmn (musée des châteaux de Malmaison et de Bois-Préau) / Franck Raux
p. 20: © The Trustees of the Natural History Museum, London
pp. 23 (left), 85: © Musée des Arts et Métiers – CNAM, Paris / photo M. Favareille
pp. 23 (right), 49, 50: © Château de Versailles, Dist. GrandPalaisRmn / Jean-Marc Manaï
p. 26: Adler Planetarium
p. 27: CCØ Paris Musées / Musée Carnavalet – Histoire de Paris
pp. 34–35: © National Maritime Museum, Greenwich, London
pp. 37, 39, 41: Paris Observatory Library
p. 44: © GrandPalaisRmn (Château de Versailles) / Michèle Bellot
p. 57: Photo: Nationalmuseum
p. 61: © MNHN – Bernard Faye
pp. 62, 68: © Château de Versailles, Dist. GrandPalaisRmn / image château de Versailles
p. 63: © GrandPalaisRmn (musée du Louvre) / Thierry Le Mage
pp. 64–65: © MNHN – Bruno Jay
pp. 67, 70: RHS Lindley Collections
p. 69: Private Collection
p. 73: © Musée du Louvre, Dist. GrandPalaisRmn / Thierry Ollivier
p. 74: © GrandPalaisRmn (Château de Versailles) / Franck Raux
p. 79: Private Collection Château de Breteuil – France
p. 82: Courtesy of The Linda Hall Library of Science, Engineering & Technology, Kansas City
p. 86: © The National Gallery, London
p. 87: © Musée Lambinet, ville de Versailles, Art Shooting
pp. 91, 92: Musée Flaubert et d'histoire de la medecine, Réunion des musées Métropole Rouen Normandie
p. 97: © GrandPalaisRmn (musée du Louvre) / Stéphane Maréchalle
p. 98 (left): Photographic credit: The Morgan Library & Museum, New York
p. 98 (right): © Musée du Louvre, Dist. GrandPalaisRmn / Pierre Philibert
p. 100: Public Domain Mark. Source: Wellcome Collection
p. 103: © Musée des Arts et Métiers – CNAM, Paris / photo S. Pelly
pp. 104–5: © Bibliothèque Mazarine
p. 106: Image courtesy The Metropolitan Museum of Art, New York
p. 107: © Musée des Arts et Métiers – CNAM, Paris / photo F. Botté
p. 116: © Victoria and Albert Museum, London

Contributors

Sarah Bond is Curator of Medicine at the Science Museum, London, where she led on two permanent galleries: *Exploring Medicine* and *Faith, Hope and Fear* (both 2019). She has previously worked for the Royal College of Nursing and Wellcome Collection. Sarah is particularly interested in the affective and sensory dimensions of material culture in relation to memory and identity.

Anne-Laure Carré holds a Master's degree from the Ecole du Louvre and a PhD in History of Technology from the Sorbonne. Since 2005 she has been the Curator of the Materials Collection at the Musée des Arts et Métiers, Paris. In 2020 she curated the exhibition *Top Modèles. Une leçon princière au XVIIIe siècle*, which focused on Madame de Genlis's models.

Stéphane Castelluccio is a Professor and Director of Research at the Centre André Chastel UMR 8150 of the Centre national de la recherche scientifique (CNRS). Specialist in the history of the royal palaces, the luxury trade and the social relationships in France during the seventeenth and eighteenth centuries, he holds three PhDs and has published more than a hundred articles and 17 books.

Hélène Delalex is Curator of Furniture and Works of Arts at the Musée national des Châteaux de Versailles et de Trianon. She has led several refurbishment projects and has curated many exhibitions including *Cheval en majesté. Au cœur d'une civilisation* (2024) and *Louis XV. Passions d'un roi* (2022–23). She was also Assistant Curator for *Sciences et curiosités à la cour de Versailles* (2010–11).

Jane Desborough is Keeper of Science Collections at the Science Museum, having worked in a curatorial capacity in the sector since 2008. She is a Liveryman of the Worshipful Company of Clockmakers and an Associate Member of the Museums Association. Her book *The Changing Face of Early Modern Time, 1550–1770* was published in 2019 by Palgrave Macmillan.

Richard Dunn is Keeper of Technologies and Engineering at the Science Museum. He was previously Curator of the History of Navigation and then Senior Curator for the History of Science at the National Maritime Museum, Greenwich. His publications include *The Telescope: A Short History* (2009), *Finding Longitude* (2014, with Rebekah Higgitt) and *Navigational Instruments* (2016).

Patricia Fara is a science historian, an Emeritus Fellow of Clare College, Cambridge, and President of the Antiquarian Horological Society. Winner of the 2022 Abraham Pais prize, she is a regular contributor to radio and TV programmes such as *In Our Time*. Her numerous books include *Science: A Four Thousand Year History* (2009) and *Life after Gravity: Isaac Newton's London Career* (2021).

Anna Ferrari is Curator of Art and Visual Culture at the Science Museum and lead curator of *Versailles: Science and Splendour* (2024–25). Since completing her PhD in the History of Art, she has curated and co-curated exhibitions at the Victoria and Albert Museum, the Royal Academy of Arts and the Sainsbury Centre for Visual Arts.

Julie Garel-Grislin is a curator at the Bibliothèque nationale de France and Head of the Curatorial Service of the Maps and Plans Department. She is interested in the history of cartographic representations and specialises in the exchange of cartographic knowledge between different mapping traditions, especially in East Asia. She is currently working on a project about imaginary maps.

Jean-François Gauvin is Professor of the History of Science and Museum Studies at the Université Laval in Québec city (Canada). He earned his PhD at Harvard University and has worked in museums and with scientific instruments for almost 30 years. His latest book is *Instruments of Knowledge. Finding Meaning in Objects, Habits, and Museums* (2023).

Matthew Howles is Associate Curator of Technologies and Engineering at the Science Museum, where he has contributed to *Science City 1550–1800: The Linbury Gallery* (2019), *Ancient Greeks: Science and Wisdom* (2021–22) and *Versailles: Science and Splendour* (2024–25). He previously worked at the Victoria and Albert Museum, having completed his Master's degree in History of Art at Bristol University.

Ludmilla Jordanova is Emeritus Professor of History and Visual Culture at Durham University. She was a Trustee of the National Portrait Gallery 2001–9 and of the Science Museum Group 2011–21. Trained in the history and philosophy of science and art history and theory, she has a long-standing interest in the history of French culture and in portraiture since 1600.

Gabriela Lamy has worked at the Palace of Versailles since 2001. She is a researcher working on the 'Plants in the Great Gardens of Europe in the Modern Era' programme led by the Research Centre of the Palace of Versailles. She is thus discovering the history of plant collections that lived in the gardens and orangeries of Versailles and Trianon.

Joséphine Lesur is a lecturer at the Muséum national d'Histoire naturelle in Paris. She is responsible for the heritage collections of Artiodactyls, Perissodactyls, Hyracoideas and Proboscideans. She is also an archaeo-zoologist and studies the relationship between humans and animals in Africa over the last 20,000 years from bone remains discovered on archaeological sites.

Katherine M. Reinhart is an Assistant Professor of Art History at Binghamton University, State University of New York. She specialises in the history of art, science and visual culture of the early modern period, focusing on the creation, use and circulation of images. Currently she is completing a monograph on the visual culture of the early Académie royale des sciences.

Béatrix Saule is a great specialist of the Palace of Versailles where she worked first as a curator and then as the museum's Director and General Curator. She curated more than a dozen international exhibitions and founded the Palace's Research Centre whose studies focus on courtly life. She curated *Sciences et curiosités à la cour de Versailles* (2010–11) and is consultant curator for *Versailles: Science and Splendour* (2024–25).

Index

References to images are in *italics*.

Published on the occasion of the exhibition *Versailles: Science and Splendour* developed with support from our expert advisor, the Palace of Versailles, and held at the Science Museum, London, 12 December 2024 to 21 April 2025. The exhibition was generously supported by Sir Sydney Lipworth KC & Lady Lipworth CBE and the Michael Marks Charitable Trust.

First published in 2024 by
Scala Arts & Heritage Publishers Ltd
43 Great Ormond Street, London WC1N 3HZ
www.scalapublishers.com
An imprint of B. T. Batsford Holdings Ltd

In association with
Science Museum Group
Exhibition Road, London SW7 2DD
www.sciencemuseum.org.uk

ISBN 978-1-78551-582-8

Project manager and copy editor: Linda Schofield
Designer: Peter Dawson, Ronja Rønning, www.gradedesign.com
Indexer: Zoe Ross
Printed in China

10 9 8 7 6 5 4 3 2

British Library Cataloguing in Publication Data.
A catalogue record for this book is available from the British Library.

Cover: Ill. 69 Print showing the demonstration of the Montgolfière balloon at Versailles on 19 September 1783, published by Le Noir, 1783.
Science Museum Group, London. Object no. 1979-527/3

Front endpaper: Design for fireworks celebrating the marriage of the comte de Provence and Marie Joséphine of Savoy, fired at Versailles on 15 May 1771, designed by firework makers Torré, Morel and Seguin, and drawn by Maurisan, 1771.
Bibliothèque municipale de Versailles. Object no. Res in-4 A 1 o

Back endpaper: Ill. 31 The Grand Ananas textile: modern recreation by Maison Pierre Frey after the original produced at the Manufacture Royale de Jouy in 1784 and owned by the Musée de la Toile de Jouy, used to redecorate Marie-Antoinette's private chambers at Versailles in 2023.

Frontispiece: Frontispiece from a suite of 21 plates showing instruments from Louis XV's cabinet of optics and physics at La Muette, engraved by Guillaume Dheulland under the direction of Dom Nicolas Noël, 1759–70.
Bibliothèque nationale de France, département des Estampes et de la photographie, Paris. Object no. IA-21-FOL

Page 8: Detail of *Veüe generale du chateau de Versailles* (General View of the Palace of Versailles), engraved by Adam Perelle, 1680s.
The Metropolitan Museum of Art, New York. Rogers Fund, 1920. Object no. 20.41(97)

Page 11: Optical microscope, mechanism by Claude-Siméon Passemant, bronze work attributed to Jacques and Philippe Caffieri, c.1750.
Musée national des châteaux de Versailles et de Trianon. Object no. V.2021.4